W9-AOS-312

Queueing Networks—
Exact Computational Algorithms

Computer Systems
Herb Schwetman, Editor

Metamodeling: A Study of Approximations in Queueing Models,
by Subhash Chandra Agrawal, 1985

Logic Testing and Design for Testability,
by Hideo Fujiwara, 1985

Performance and Evaluation of LISP Systems,
by Richard P. Gabriel, 1985

The LOCUS Distributed System Architecture,
edited by Gerald Popek and Bruce J. Walker, 1986

Analysis of Polling Systems,
by Hideaki Takagi, 1986

Performance Analysis of Multiple Access Protocols,
by Shuji Tasaka, 1986

Performance Models of Multiprocessor Systems,
by M. Ajmone Marsan, G. Balbo, and G. Conte, 1986

A Commonsense Approach to the Theory of Error Correcting Codes,
by Benjamin Arazi, 1987

Microprogrammable Parallel Computer: MUNAP and Its Applications,
by Takanobu Baba, 1987

Simulating Computer Systems: Techniques & Tools,
by M. H. MacDougall, 1987

Research Directions in Object-Oriented Programming,
edited by Bruce Shriver and Peter Wegner, 1987

Object-Oriented Concurrent Programming,
edited by Akinori Yonezawa and Mario Tokoro, 1987

Networks and Distributed Computation,
by Michel Raynal, 1988

Queueing Networks - Exact Computational Algorithms: A Unified Theory Based on Decomposition and Aggregation,
by Adrian E. Conway and Nicolas D. Georganas, 1989

Fault Tolerance Through Reconfiguration in VLSI and WSI Arrays,
by R. Negrini, M. G. Sami, and R. Stefanelli, 1989

Queueing Networks—Exact Computational Algorithms

A Unified Theory Based on Decomposition and Aggregation

Adrian E. Conway and Nicolas D. Georganas

The MIT Press
Cambridge, Massachusetts
London, England

This book was printed and bound in the United States of America.

Library of Congress Catologing-in-Publication Data

Conway, Adrain E.
Queueing networks—exact computational algorithms.

(Computer systems)
Bibliography: p.
Includes index.
1. Computer networks. 2. Queuing theory. I. Georganas, Nicolas D. II. Title. III. Series: Computer systems (Cambridge, Mass.)
TK5105.5.C668 1989 004.2'1 89-12556
ISBN 0-262-03145-0

To our parents, who gave us light

CONTENTS

List of Tables

List of Figures

SERIES FOREWORD

This series is devoted to all aspects of computer systems. This means that subjects ranging from circuit components and microprocessors to architecture to supercomputers and systems programming will be appropriate. Analysis of systems will be important as well. System theories are developing, theories that permit deeper understandings of complex interrelationships and their effects on performance, reliability, and usefulness.

We expect to offer books that not only develop new material but also describe projects and systems. In addition to understanding concepts, we need to benefit from the decision making that goes into actual development projects; selection from various alternatives can be crucial to success. We are soliciting contributions in which several aspects of systems are classified and compared. A better understanding of both the similarities and the differences found in systems is needed.

It is an exciting time in the area of computer systems. New technologies mean that architectures that were at one time interesting but not feasible are now feasible. Better software engineering means that we can consider several software alternatives, instead of "more of the same old thing," in terms of operating systems and system software. Faster and cheaper communications mean that intercomponent distances are less important. We hope that this series contributes to this excitement in the area of computer systems by chronicling past achievements and publicizing new concepts. The format allows publication of lengthy presentations that are of interest to a select readership.

Herb Schwetman

PREFACE

This book provides a state-of-the-art view of the field of exact computational algorithms for the analysis of product-form queueing networks in equilibrium. Recent developments that have been made in the field, and more established results, are presented within the context of a unified constructive theory based on the general notions of decomposition and aggregation. The theory provides much intuitive insight into the general problems of constructing efficient computational algorithms and parametric analysis techniques. Furthermore, it constitutes a basis for the possible future development of new algorithms with improved efficiency. The inherent generality of the constructive theory suggests, in fact, the existence of a range of new algorithms that might possibly be developed.

The book will be valuable to those who are carrying out fundamental analytical research in the areas of computer system and computer network performance evaluation. It will also be useful as an up-to-date reference for mathematical background material on product-form queueing networks and for the implementation of the various exact algorithms that now exist.

The perspectives that we provide in this book have their origin in the doctoral research carried out by A.E. Conway at the Electrical Engineering Department of the University of Ottawa, Ottawa, Canada, and at the Advanced Studies Department of Bull-Transac, Paris, France. The Recursion by Chain Algorithm (RECAL) and the notion of Decomposition by Chain constituted a part of the Ph.D. dissertation of A.E. Conway. The origins of the unified constructive theory for exact computational algorithms and the conjecture concerning the existence of an optimal algorithm are also contained in this dissertation.

A portion of this book was written during the time that A.E. Conway was a Visiting Assistant Professor in the Department of Electrical Engineering of McGill University, Montréal, Canada. The support provided by the Department of Electrical Engineering of McGill is gratefully acknowledged. The preparation of this book was made using the facilities of GTE Laboratories Incorporated. The authors are grateful for this support that facilitated the production of a camera-ready copy. The authors are also grateful to Dr. Brian E. Conway for help in proofreading and to Regina Sutton for assistance in text processing. Finally, A.E. Conway is grateful to Michel Ouellet for his encouragement.

A.E.C.
N.D.G.

Boston and Ottawa, September 13, 1988

Queueing Networks—
Exact Computational Algorithms

CHAPTER 1

Preliminaries

1.1 Introduction

In the last fifteen years, queueing networks have come into widespread use as stochastic models of computer systems and computer communication networks. The fundamental common feature of these systems is that there is a limited resource which must be shared among a number of competing customers who require service. Within the context of computer systems, the customer takes on the form of a job or task and the resource consists of CPUs, memory and I/O devices. In computer networks, the customer is a message or packet and the resource consists of the data transmission channels, processors and buffers of the underlying data communication network. Queueing network modeling provides a discipline for the analysis of systems which are stochastic in nature and which involve a flow of traffic in a network of resources.

An important factor in the adoption of queueing networks as mathematical models of systems which may be characterized as a network of stochastic service centers is that experience has shown that the performance predictions obtained often agree quite well with the performance characteristics of actual systems [SPR1]. Although it cannot be proved in general, queueing network models appear to be fairly robust [SUR1], despite the simplifying assumptions that are necessarily made, as in any mathematical model which is an abstraction of reality.

The importance of queueing network modeling as a tool for system design and performance evaluation appears to be increasing. With the ongoing technological advances that are being made, there is the feasibility of implementing more sophisticated systems.

Examples include distributed computing systems and computer communication networks with higher degrees of functionality. In these systems there exist many important queueing theoretical design issues which remain, at present, unresolved. There are great potential benefits to be gained with the application of quantitative design methods based on queueing network analysis.

There are two main approaches that exist for the analysis of queueing network models. The method of wider applicability is Monte Carlo simulation [HAM1] since, by nature, it enables the modeling and analysis of mechanisms which may be mathematically intractable. A limitation of the simulation method is, however, the potentially high computational cost to obtain reasonable confidence intervals on system performance measures. Furthermore, it is obvious that the method is incapable of providing any general mathematical results which characterize the behaviour of a system.

The other main approach is the analytical method. Analytic results can take on the form of a closed-form solution or equations of a functional type which are solved using a computational algorithm. Both of these forms can provide much insight into the manner in which various parameters affect the performance of a system. The advantage of following the analytical method is that the results are of general applicability, under the specified mathematical assumptions, and the computational costs are usually significantly lower than those of the simulation method. A disadvantage of the analytical method is that the computational algorithms themselves may have excessive computational requirements or a mathematical solution of a closed or functional type may simply not be known.

As a result of the inherent limitations of both methods of analysis, an approach which has become more common today in practice is the adoption of hybrid methods that attempt to combine simulation with analytical results [KUR1]. There has also been an increase in the application of analytical approximation methods, such as bounding algorithms [EAG1,EAG2] and asymptotic expansion techniques [MCK1,MCK4,RAM1], and heuristic type 'analytical' methods [CHA2,MAR1,NEU1,NEU2,SOU3,SOU4,SOU5] that are based, to a certain extent, on exact analytical results. All of these approaches are commonly regarded as analytical performance evaluation methods since they all involve, to varying degrees, the application of mathematical analysis. Although the heuristic methods may not always have an entirely sound mathematical basis, empirical evidence has shown that they often provide satisfactory results.

Within the broad category of analytical methods for the analysis of queueing networks, we may differentiate between specialized techniques that presume a certain predetermined number of service centers or a particular network configuration, such as central-server, tandem or cyclic, and techniques which are applicable to networks of queues in which such *a priori* constraints are not assumed. The methods of general applicability, that do not involve any use of simulation, include the following:

(1) Matrix methods for Markovian queueing networks

(2) Exact computational algorithms for product-form queueing networks

(3) Approximation methods for large product-form queueing networks

(4) Heuristic methods for large product-form queueing networks

(5) Heuristic methods for general non-product-form queueing networks.

This book is primarily concerned with developments that have been made in the second category, namely, in general purpose exact solution algorithms for the known class of queueing networks whose state-probability distribution has the form of a product of terms.

The area of exact computational algorithms for product-form queueing networks may be viewed as being one of the cornerstones of queueing network analysis. Queueing networks of the product-form type remain the only broad category of networks for which general analytical results are known concerning the probability distribution of random variables of interest. They are widely used as models and there is a constant need for efficient computational algorithms that can be applied to this important class of networks. Furthermore, the exact algorithms that have been developed in the past have provided a formal theoretical foundation for the development of a wide range of the commonly used approximation and heuristic methods for large product-form networks and heuristic methods for general queueing networks that do not support a product-form solution. It is likely that any progress made in the area of exact algorithms may possibly lead to further developments in the other related areas.

The primary purpose of this book is to provide a state-of-the-art view of the field of exact computational algorithms for product-form queueing networks and expose some perspectives we have on the general problem of constructing efficient algorithms. Significant recent developments, including the *Recursion by Chain Algorithm* (RECAL) of Conway and Georganas [CON3], the *Mean Value Analysis by Chain* (MVAC) algorithm of Conway, de Souza e Silva and Lavenberg [CON2,CON6,CON9,SOU1] and the *Distribution Analysis by Chain* (DAC) algorithm of de Souza e Silva and Lavenberg [SOU2], and more established algorithms, including the *Convolution Algorithm* of Buzen [BUZ1,BUZ2] and of Reiser and Kobayashi [REI3,REI8] and the *Mean Value Analysis* (MVA) algorithm of Reiser and Lavenberg [REI4], are presented in a unified fashion. Rather than simply providing a compendium of all of the exact algorithms that have been developed to date, our approach in this book is to first develop a general unified theory for the construction of exact computational algorithms. This serves to place us, from the very outset, at a theoretical vantage point from which we may view the developments that have been made and see a close connection between them.

The proposed general methodology for the construction of computational algorithms, is based on the notions of decomposition and aggregation and on certain results from the theory of reversible Markov chains. It is constructive in nature and utilizes probabilistic concepts. As a consequence, it is pedagogically appealing and provides much intuitive insight into the problem of constructing an efficient algorithm. The proposed general methodology serves to unify the perhaps seemingly unconnected developments that have been made in a way which is intuitively appealing. The general theory actually defines a whole class of exact computational algorithms which could, in principle, be developed and includes, as special cases, the main exact algorithms that have been developed to date. The theory suggests the existence of other efficient computational algorithms which could possibly be developed and which may be even more efficient than those known today. This conjecture may provide a catalyst for future progress to be made in the area.

Apart from providing a unified presentation of the developments that have taken place in recent years and a general methodology for the construction of efficient algorithms, this book is also intended to serve as an up-to-date reference source for the main known theoretical results pertaining to product-form queueing networks in equilibrium and for the implementation of the various exact computational algorithms. The discussions pertaining to the general methodology and the unifications are restricted to Chapters 3 and 4 and will primarily be of interest to those who are pursuing fundamental research in the area. The theoretical results, of Chapter 2, on product-form queueing networks, and the survey of computational algorithms in Chapter 5 will primarily be of interest to those who wish to acquire an up-to-date familiarity with the main results in the area or implement the algorithms to solve certain performance evaluation problems. The book is structured so that Chapters 2 and 5 may be read together on their own.

The position adopted in this book is that there exists a unified theoretical framework, based on the general theory of decomposition and aggregation, within which we may view the general problem of constructing exact computational algorithms. The following section overviews and explains the sequence of developments to be made in the book that lead to this unified theory. These developments are related to previous work in the area. In the following section, we also survey the main developments that have taken place over the years in the area of exact algorithms.

1.2 Overview

Given a queueing network of the product-form type and its associated parameters such as the routing probabilities and the service-time distributions of the customers at the various service centers, the main objective of an exact computational algorithm is to produce, in a computationally efficient manner, the mean performance measures of interest, such as the mean queue-lengths, mean waiting-times and throughputs of the customers at the queues. From the point of view of one who is carrying out system performance evaluation studies, the exact computational algorithm may be viewed as a 'black box' which maps the parameters of a product-form queueing network into the mean performance measures. An overriding practical concern is the computational cost involved in this transformation.

The origins of exact computational algorithms for product-form queueing networks may be attributed [REI7] independently to Buzen [BUZ1,BUZ2] and to Reiser and Kobayashi [REI8] who, in the early 1970's, developed an efficient algorithm for the evaluation of the mean performance measures of product-form queueing networks of the Gordon-Newell [GOR1] type. The intended application was to time-sharing models of computer systems. Buzen, and Reiser and Kobayashi, developed an efficient procedure for the computation of the normalization constant which is associated with the equilibrium state-distribution of the network and derived formulae for the mean performance measures of interest in terms of normalization constants. The algorithm of Buzen, and of Reiser and Kobayashi, is now commonly known as the *Convolution Algorithm* since the underlying recursive difference equation, which is utilized, is of the form of a discrete convolution operation.

The Gordon-Newell queueing network assumes a single class of customers in which each customer is indistinguishable with respect to the routing-parameters and service-time requirements at the various service centers. Shortly after the development of the original Convolution Algorithm, the Gordon-Newell network was extended significantly to accommodate multiple closed classes of customers and open classes, in which there may be exogeneous arrivals into the network. These extensions may be attributed to Baskett et al [BAS1], Kelly [KEL1,KEL2] and Reiser and Kobayashi [REI3]. The extension of the Convolution Algorithm to networks with multiple closed classes of customers and open classes was made by Reiser and Kobayashi [REI3] in 1975. The extended algorithm is essentially a generalization of the original Convolution Algorithm to a multi-dimensional convolution operation. Reiser and Kobayashi interpreted their algorithm within the context of multi-dimensional digital filtering theory [RAB1]. The work of Reiser and Kobayashi greatly extended the domain of applicability of queueing networks as models. Multiple classes of customers enabled the modeling of computer communication networks with multiple virtual channels under window flow control and the modeling of computer systems with heterogeneous classes of jobs, including batch and time-sharing classes.

Following the development of the generalized Convolution Algorithm of Reiser and Kobayashi, a new approach was formulated for the computation of the mean performance measures of product-form queueing networks. In the late 1970s, Reiser and Lavenberg [REI4] developed the so-called *Mean Value Analysis* (MVA) theory for the analysis of multiple class closed queueing networks. The striking feature of the new approach was that the recursive equations used for computing the measures of interest were themselves entirely in terms of mean performance measures and did not involve any use of normalization constants. The theoretical basis for this new type of recursion was provided by the then recently discovered Arrival Theorem [LAV1,SEV1] for product-form queueing networks. The development of the MVA algorithm did much to popularize the use of multiple class queueing networks as models. The algorithm has proven to be easier to understand and simpler to implement than the Convolution Algorithm. It has also proved to be pedagogically appealing since it is amenable to a probabilistic interpretation. The probabilistic insight provided by the MVA theory also led to the development of a range of iterative (e.g. [AKY1,CHA2]) and recursive heuristic techniques (e.g. [BRY1,SOU4]). These have enabled the approximate analysis of large product-form queueing networks and of queueing networks which do not support a product-form state-distribution.

Since the development of the MVA theory, a number of other exact computational algorithms have been developed. Notable among these are specialized algorithms that exploit sparsity or locality that is often found in the routing of customers in a queueing network model. These algorithms include the *Tree Convolution* algorithm of Lam and Lien [LAM3] and the *Tree MVA* algorithms of Hoyme et al [HOY1] and Tucci and Sauer [TUC1]. All of these algorithms are closely related to their respective general versions. These algorithms have enabled the solution of large computer communication network models with many window flow controlled virtual channels.

Recently, further fundamental progress has been made in the area of exact computational algorithms for queueing networks with multiple closed classes of customers. An important development which has appeared is the *Recursion by Chain Algorithm* (RECAL) of Conway and Georganas [CON1,CON3]. The recursion employed in this scheme is based on the notion of *'Decomposition by Chain'*

[CON1,CON4], to be described in Chapter 3, and is significantly different in structure from the recursions employed in the Convolution and MVA algorithms. As in the Convolution Algorithm, RECAL involves the computation of certain normalization constants but the manner in which this is carried out is completely different. The main feature of RECAL is that the growth of the computational costs involved in obtaining the mean performance measures is a polynomial function in the number of distinct closed classes of customers, with the degree of the polynomial growing linearly in the number of service centers. In the Convolution Algorithm and MVA, the costs grow exponentially in the number of distinct closed classes. The great difference between the computational growth characteristics of these algorithms raises some interesting theoretical questions concerning the inherent computational complexity of product-form queueing network problems.

Another fundamental development of importance which has been made recently in the area of exact computational algorithms is the *Mean Value Analysis by Chain* (MVAC) algorithm of Conway, de Souza e Silva and Lavenberg [CON2,CON6,CON9,SOU1]. As in the original MVA algorithm, the underlying recursive equations employed in MVAC are themselves expressed in terms of mean performance measures and have a simple probabilistic interpretation. The recursive structure of MVAC is closely related to that of RECAL and both MVAC and RECAL have essentially the same computational costs. The main advantage of MVAC over RECAL is its inherent numerical stability. In RECAL, the computation of the normalization constant has the potential for numerical overflow or underflow. This same potential problem is known to exist in the Convolution Algorithm [LAM1].

A final recent development of fundamental importance in the area of exact algorithms is the *Distribution Analysis by Chain* (DAC) algorithm of de Souza e Silva and Lavenberg [SOU2]. This algorithm was primarily developed for the computation of the *joint* marginal queue-length distribution of product-form queueing networks with multiple closed classes of customers. One particular application of importance where this algorithm is found to be of practical use is in the analysis of computer system availability models (to be considered in Subsection 2.3.8), where joint queue-length distributions are often required to determine the availability measure. The DAC algorithm may also be employed to obtain the mean performance measures of product-form queueing networks in general. The DAC algorithm is, therefore, another general purpose exact algorithm. The computational costs of DAC are very similar to those of RECAL and MVAC. Since the underlying recursive equations on which the DAC algorithm is based are in terms of joint queue-length probabilities rather than normalization constants, the DAC algorithm may be classified as another type of MVA algorithm.

Looking back at the principal developments that have been made in the last decade and a half, we may recognize a number of different exact algorithms with very different theoretical foundations. Table 1.1 summarizes the general features that the various algorithms have in common. The theoretical relationship

	Computational Costs Exponential in R	Computational Costs Polynomial in R
Normalization Constant Approach	Convolution	RECAL
Mean Value Approach	MVA	MVAC DAC

R = Number of Closed Routing Chains

Table 1.1
Classification of Exact Computational Algorithms

between RECAL and MVAC is apparent since both are based on the notion of Decomposition by Chain. It has also been shown by Lam [LAM2] that there exists a close connection between the Convolution Algorithm and MVA, in the sense that each may be derived algebraically from the other. A connection between RECAL and MVAC, on the one hand, and the Convolution Algorithm and MVA, on the other, is at first sight not apparent. Furthermore, it is not apparent that the DAC algorithm is related directly to any of the other exact algorithms. A general theory for the construction of computational algorithms, which can unify, at a theoretical level, all of the developments that have been made, remains to be defined. One of the purposes of this book is to present the main developments described above within the context of a unified theory based on the general theory of decomposition and aggregation, as originally developed by Simon and Ando [SIM1] in 1961.

In its original form, the decomposition and aggregation methodology of Simon and Ando is applicable to Markov chains in general and can be used to reduce the problem of obtaining the equilibrium state-distribution into a set of smaller problems. In general, the method does not yield exact results. However, it is naturally suited to the analysis of systems which satisfy certain technical conditions of *near-complete decomposability* in which the transient state-distribution of the system is known to evolve through certain phases of short-term dynamics, short-term equilibrium, long-term dynamics and, finally, long-term equilibrium. Under these conditions, which can be defined rigorously, it has been shown by Courtois [COU1,COU3] that the decomposition and aggregation procedure yields results with an error of the order of ε, where ε is the so-called '*maximum degree of coupling*' between the subsystems involved in the decomposition. As the degree of coupling tends to zero, so does the error incurred by the decomposition and aggregation procedure. Courtois also established sufficient conditions for a Markov chain to satisfy the required technical conditions of near-complete decomposability [COU1].

Courtois [COU8] was the first to apply the general theory of Simon and Ando to the problem of analyzing queueing networks arising within the context of the modeling of computer systems. Courtois considered a Gordon-Newell queueing network with N service centers and applied the general theory of Simon and Ando to decompose the problem of obtaining the equilibrium state-distribution into the analysis of subsystems that could themselves be characterized as smaller queueing networks, each having (N-1) service centers. Each of these smaller queueing networks could, in turn, be decomposed in a similar recursive manner. This hierarchic method of analysis can be termed decomposition and aggregation by service center, or simply, 'Decomposition by Service Center'. Courtois obtained sufficient conditions for a Gordon-Newell network to be nearly-completely decomposable by service center. More relaxed sufficient conditions have subsequently been reported in [BALS1].

Subsequent to the work of Courtois, Vantilborgh [VAN1] found necessary and sufficient conditions under which the Decomposition by Service Center approach of Courtois could be carried out with no error. Courtois [COU2] later established the necessary and sufficient conditions under which the Simon and Ando technique is guaranteed to yield exact results for Markov chains in general. By applying these general conditions, Courtois [COU2] showed how the exact procedure of Vantilborgh could be extended to product-form queueing networks with multiple closed classes of customers, of the type considered by Baskett et al [BAS1].

The work of Courtois and Vantilborgh has served to show that, for product-form queueing networks, exact results can always be obtained by the decomposition and aggregation procedure under the presumption that the network is decomposed by service center. It has long been recognized, however, that, relative to existing computational algorithms such as the Convolution Algorithm, the direct application of the Decomposition by Service Center technique to product-form queueing networks is an inefficient procedure for obtaining the mean performance measures of interest. This was clearly pointed out by Courtois in [COU1, Section 6.4]. Rather, it has been recognized that the main use of the decomposition and aggregation methodology for product-form queueing networks is "*...to essentially unearth the basic relations that exist among levels of aggregation and among aggregate resources in networks of stochastic service systems*" [COU1, page 87]. These relations can provide much intuitive insight into the structure of a queueing network problem.

Although the Decomposition by Service Center method of Courtois has not proved to be of direct practical use in providing a basis for any new *exact* efficient computational algorithms for product-form queueing networks, it did lay the foundation for the development of many heuristic approximation algorithms for general queueing networks that do not support a product-form state-distribution (e.g. [BRA1,CHA4,MAR3,NEU1,NEU2,SCH1,VAN2]). These approximation algorithms, which exploit directly the notion of Decomposition by Service Center, have proved to be of much practical use. It may be observed, however, that the Decomposition by Service Center approach of Courtois is only one particular method of decomposing a queueing network problem. In the abstract, a product-form queueing network is simply a continuous-time Markov chain with a particular structure and the state-space of this Markov chain can be partitioned in a number of distinct ways. Each of these possible state-space partitions can be associated with a certain subsystem, the solution of which can be identified with the problem of analyzing a particular queueing network problem. Apart from the newly developed Decomposition by Chain [CON4] technique to be considered in Chapter 3 and a special method of decomposition described in [ZAH1] to deal with high service-time variabilitics, it seems from the literature that no other attempts have been made to exploit other methods of decomposition that may possibly be of practical use. The proliferation of results in the area, that make use of the notion of Decomposition by Service Center, attests to the wide influence that the work of Courtois has had in the field. A result which we shall establish, and one which is used as a basis for our constructive theory, is that it is possible to decompose a product-form queueing network *arbitrarily* and be assured that exact results will always follow in the application of the general Simon and Ando decomposition and aggregation procedure. This observation generalizes the exact Decomposition by Service Center methods of Courtois and Vantilborgh and, as will be seen, makes the proof almost immediate that exact results may be obtained by these methods.

Having established that it is possible to arbitrarily decompose a product-form queueing network into a hierarchy of interrelated subsystems and maintain exact results, a generalized decomposition and aggregation procedure for product-form queueing networks may be formulated which includes the Decomposition by Service Center and the Decomposition by Chain methods as special case. This generalized decomposition defines a general hierarchical recursive structure to be found in a product-form queueing network. As with the Decomposition by Service Center method of Courtois, however, an analysis by decomposition and aggregation based directly on this general recursive structure would, in general, prove to be inefficient relative to existing computational algorithms. Nevertheless, in the generalized decomposition, it is possible to identify subsystems at each level in the hierarchy which may be formally identical, as far as their state-distributions are concerned. Such subsystems, with equivalent state-distributions, were indeed identified by Courtois in the method of decomposition that he adopted. Courtois recognized that, since they had identical distributions, the complexity of the general decomposition and aggregation procedure could be reduced. It is this existence of equivalence classes of subsystems, or redundancy, at the various levels in our generalized hierarchical decomposition which is exploited to formulate a general efficient recursive structure that can be used as the basis for the construction of efficient computational algorithms.

The proposed general methodology for the construction of efficient computational algorithms is based directly on the observations that we have made above and may be summarized as follows. The initial step in the method is to adopt some particular hierarchical method of decomposition. As has been mentioned above, the general decomposition and aggregation procedure will yield exact results for any decomposition we may care to adopt in a product-form queueing network. The actual choice of a network decomposition is left completely open to us and the efficiency of the resulting algorithm depends completely on the choice made in this step. The second step is to identify, at each level in the adopted hierarchical decomposition, the equivalence classes of subsystems, if any. The number of equivalence classes at each level, and the number of levels that make up the hierarchy, are the factors that determine the efficiency of the resulting algorithm. These factors, however, are completely dependent on the choice of decomposition that is initially made. The final step is to determine the recursive relationships that exist between the subsytems at neighboring levels in the decomposition. Such relationships may take on the form of recursive equations involving, for example, the normalization constants, the marginal queue-length distributions or the mean performance measures associated with the subsystems. Such relationships may be found since, according to the general decomposition and aggregation theory of Simon and Ando, there always exists a relationship that can be established between the state-distributions of the subsystems at adjacent levels. Working directly in terms of these state-distributions is, however, to be avoided since this, in general, would prove to be very inefficient. Having carried out the above summarized construction, one may then attempt to algebraically simplify the recursive equations obtained and analyze the computational requirements of the resulting recursive algorithm. These requirements include the number of operations and the storage space requirements. These two requirements are usually closely dependent on each other. Sometimes there exist opportunities for certain space/time tradeoffs to be made.

In subsequent chapters, we will show how the Convolution Algorithm, MVA, RECAL, MVAC and DAC may each be constructed according to the above described steps. This serves to unify the developments, referred to above, that have been made. It is the inherent generality of the proposed methodology which suggests the existence of other algorithms that could possibly be developed. For any decomposition we may care to define in the initial step, there results an algorithm having certain computational costs. In fact, for a given queueing network with some number of service centers and closed classes of customers, there exists an algorithm which is computationally optimal within the class of algorithms defined by our methodology.

Apart from providing a constructive method for derivation of computational algorithms, the observations we have made above can also be used as a basis for the derivation of generalized methods of parametric analysis. Such analysis is useful when one is interested in varying a certain parameter, or a set of parameters, and observing the effect on network performance measures without having to analyze repetitively the entire queueing network under consideration. In the literature, there currently exist two known methods of parametric analysis. The first method, originally developed by Chandy et al [CHA1], and commonly known as *Norton's Theorem* for queueing networks, can be used for the parametric analysis of a network with respect to the parameters that are local to a certain service center. The other method, known as *Parametric Analysis by Chain*, which has recently been developed by Conway and Georganas [CON4], can be used for the parametric analysis of a network with respect to the routing-parameters of a particular closed class of customers. As will be seen, the problem of parametric analysis is intimately related to the problem of constructing a computational algorithm. The fundamental difference between the construction of a computational algorithm and the development of a parametric analysis technique, using the methodology that we have described, is that, in the former, we adopt a multiple-level hierarchical decomposition while, in the latter, we use a single-level decomposition that isolates the portion of the network that contains the parameters of interest. Since a product-form queueing network may be decomposed arbitrarily, with the assurance that exact results will follow in an analysis by decomposition and aggregation, it is possible to isolate, by suitable choice of decomposition, any parameters that may be of interest and construct an exact method of parametric analysis based on the relationships that can be established between the subsystems specified by the decomposition which has been adopted. The resulting generalized method of parametric analysis includes the known parametric methods as special cases and suggests the existence of other parametric methods that may be of practical use for certain types of performance evaluation studies. The notion of generalized parametric analysis is treated in Chapter 3, prior to considering the problem of constructing computational algorithms. The development of the constructive theory for computational algorithms follows quite naturally as an extension of the theory developed for generalized parametric analysis.

The sequence of developments, to be made in subsequent chapters, may be summarized as follows. In the second chapter, we summarize the main theoretical results of practical importance pertaining to product-form queueing networks and overview a number of current applications. In the third chapter, we begin by describing the general decomposition and aggregation theory of Simon and Ando and the main results connected with it. The general conditions are reviewed under which the decomposition and aggregation procedure yields exact results. In Section 3.2, within the context of the Simon and Ando theory, we consider the method of Decomposition by Service Center of Courtois, as applied to product-form queueing networks. In Section 3.3, we show how this decomposition method provides a basis for the exact parametric analysis of a queueing network with respect to the parameters of a particular service center of interest. We also show that the resulting parametric analysis technique is equivalent entirely to the Norton's Theorem for queueing networks. In Section 3.4, we consider the Decomposition by Chain procedure and, in Section 3.5, show how it provides a basis for exact parametric analysis with respect to a particular closed class of customers of interest. In Section 3.6, using a simple argument, we prove that the general decomposition and aggregation procedure of Simon and Ando always yields exact results both for reversible [KEL3] Markov chains and reversible product-form queueing networks. As is to be explained in Section 3.6, the condition of reversibility is one which does not impose any special restrictions on a product-form queueing network since it can always be imposed without altering the equilibrium state-distribution. This result provides the basis for methods of parametric analysis that may be applied to arbitrary subsystems of interest in product-form queueing networks. This generalized method of parametric analysis is described in Section 3.7.

The development of our general theory for the construction of efficient computational algorithms, is made in Chapter 4. In Section 4.1, we formalize a generalized decomposition and aggregation procedure for multiple-chain product-form queueing networks. In Section 4.2, we show that, having done this, we may then identify equivalence classes of subsystems in the various levels of the hierarchy. This then allows us to define a generalized recursive structure to be found in a queueing network which may be used as a formal basis for the construction of exact computational algorithms. The unified methodology for the construction of computational algorithms is formalized in Section 4.3. In Section 4.4, we show how it is possible to derive the Convolution Algorithm within this generalized framework when the adopted network decomposition is by service center. In Section 4.5, we show how the recursion of the MVA algorithm may be derived when the network decomposition is based on the position in the network of a particular customer. In Sections 4.6 and 4.7, respectively, we show how the recursions of RECAL and MVAC may be derived when the adopted network decomposition is by routing chain. Finally, in Section 4.8, we show how the recursion of the DAC algorithm may be derived when, as for the MVA algorithm, the network decomposition is based on the position in the network of a particular customer. These results serve to unify all of these algorithms within a general framework. As mentioned above, the inherent generality of our proposed methodology suggests the existence of other efficient algorithms that could possibly be developed. Such possibilities are considered in Sections 4.9 and 4.10.

The final chapter is to serve as a basis of reference for the main algorithms that have been developed up until now. In Sections 5.1 to 5.5, respectively, we consider the details of the Convolution Algorithm, MVA, RECAL, MVAC and DAC. Finally, in Section 5.6, we consider specialized algorithms which exploit structure that may be found in certain types of queueing network models.

Bibliography

D. Ferrari, *Computer Systems Performance Evaluation*, Prentice-Hall, Englewood Cliffs, New Jersey, 1978.

E. Gelenbe, and I. Mitrani, *Analysis and Synthesis of Computer Systems,* Academic Press, London, England, 1980.

J.F. Hayes, *Modeling and Analysis of Computer Communications Networks*, Plenum Press, New York, 1984.

H. Kobayashi, *Modeling and Analysis: An Introduction to System Performance Evaluation Methodology*, Addison-Wesley, Reading, Massachusetts, 1978.

S.S. Lavenberg (Ed.), *Computer Performance Modeling Handbook*, Academic Press, New York, 1983.

E.D. Lazowska, J. Zahorjan, G.S. Graham, and K.C. Sevcik, *Quantitative System Performance: Computer System Analysis Using Queueing Network Models*, Prentice-Hall, Englewood Cliffs, New Jersey, 1984.

C.H. Sauer, and K.M. Chandy, *Computer Systems Performance Modeling,* Prentice-Hall, Englewood Cliffs, New Jersey, 1981.

CHAPTER 2

Theory and Applications of Queueing Networks

This chapter provides an introduction to queueing networks and a summary of the main modeling features and theoretical results on the equilibrium state-distribution of product-form queueing networks. Use will be made of these results in the developments described in subsequent chapters. The chapter also illustrates, by means of a number of examples, the wide range of current applications of product-form queueing network models. Survey papers on equilibrium results for queueing networks include [DIS1,LEM1].

2.1 Equilibrium Results for Queueing Networks

A network of queues consists of a set of service centers arranged at the nodes of a graph. Customers travel along the edges of the graph and receive service at the various nodes. A queueing network is *closed* if there are no exogeneous arrivals of customers and no departures from the network. A closed network contains a fixed population of customers that circulate continuously among the service centers. A network is *open* if customers may enter and eventually leave the network. The population of an open network is unconstrained. A network is *mixed* if for certain customers the network is closed while for others it is open.

The description of a queueing network includes the specification of the mechanism whereby customers arrive at and depart from the network, of the manner in which the customers are routed between the nodes, of the queueing disciplines at the service centers and of the service-requirements. If, for example, it is assumed that the arrival processes, if any, are Poisson, that the routing is random and that the service-time requirements are exponentially distributed random variables, then the queueing network will be Markovian. Under such circumstances, which imply that there is no memory in the system, the analysis of a queueing network is reduced to the problem of analyzing a Markov chain in equilibrium. If $\mathbf{Q}$ is the infinitesimal generator for the continuous-time Markov chain associated with the queueing network and π is the equilibrium state-distribution, then π may be found by solving the set of simultaneous linear 'global balance' equations

$$\pi \mathbf{Q} = \mathbf{0}$$

subject to the constraint that

$$\pi \mathbf{1}^T = 1,$$

where $\mathbf{1}^T$ is a compatible column vector, all of whose elements are unity, and $\mathbf{0}$ is a row vector of zeros. The measures of interest in a queueing network, such as the mean queue-lengths and throughputs, may in principle be calculated having obtained π. In queueing networks, however, the state-space is infinite or usually extremely large. The state variable is multi-dimensional in nature, due to the presence of a set of service centers, and the state description may also be augmented by the inclusion of the description of the ordering of the customers in the queues. Furthermore, the service and routing mechanisms may, in certain circumstances, be quite complex. These factors, in general, preclude the analysis of infinite state-space or finite state-space Markovian queueing networks by purely analytical methods and the numerical analysis of finite state-space queueing networks by direct matrix methods [STE1].

Over the years, however, a general class of queueing networks has been discovered for which the form of the equilibrium distribution π is known. This is the so-called class of product-form queueing networks which we are concerned with here. In these networks, the distribution π has the general form of a product of terms. The origin of product-form queueing networks may be attributed to Jackson [JAC1] who first considered an open network of exponential queues, each of the first-come first-served (FCFS) multi-server type. Subsequent generalizations to the classical Jackson network have been made over the years. Today, there are a large variety of general modeling features available within the class of product-form queueing networks that may be applied to performance evaluation problems. In the following, we consider the original Jackson network and the main extensions that have subsequently been made.

2.1.1 The Jackson Queueing Network

A Jackson queueing network, consisting of N service centers of the $M/M/K/\infty$ type, is illustrated in Figure 2.1. The network is open and customers from the outside arrive at the network according to a Poisson process. The rate of exogeneous arrivals to node i, $i = 1,...,N$, is λ_i. The service-time requirement distributions for the customers at the various service centers are assumed to be exponential and independent of one another. The mean service-requirement for a customer at node i is t_i. At node i, it is assumed that there are k_i parallel servers that each accomplish work at unit rate. The routing of the customers between the nodes of the network is a random walk. A customer who completes service at node i proceeds next to node j with probability p_{ij} or departs the network with probability p_{i0}, where

$$p_{i0} = 1 - \sum_{j=1}^{N} p_{ij}.$$

The state of the network is $\mathbf{n} = (n_1,...,n_N)$, where n_i is the total number of customers at node i. The state-space is S, where

$S = \{\mathbf{n} \mid \mathbf{n} = (n_1,\ldots,n_N),\ n_i \geq 0,\ i = 1,\ldots,N\}.$

For the Jackson queueing network, the explicit equilibrium balance equation for the state-probability $\pi(\mathbf{n})$, $\mathbf{n} \in S$, is

$$\pi(\mathbf{n}) \sum_{i=1}^{N} (\lambda_i+\beta_i(n_i)(1-p_{ii})) = \sum_{i=1}^{N}\sum_{j=1}^{N} \pi(\mathbf{n}+\mathbf{1}_i-\mathbf{1}_j)\beta_i(n_i+1)p_{ij}\delta(\mathbf{n}+\mathbf{1}_i-\mathbf{1}_j)\delta(i,j)$$

$$+ \sum_{i=1}^{N} \pi(\mathbf{n}-\mathbf{1}_i)\lambda_i\delta(\mathbf{n}-\mathbf{1}_i) + \sum_{i=1}^{N} \pi(\mathbf{n}+\mathbf{1}_i)\beta_i(n_i+1)p_{i0}, \tag{2.1}$$

where $\delta(\mathbf{n}) = 1$, if $\mathbf{n} \in S$, and 0 otherwise, $\delta(i,j) = 0$, if $i = j$, and 1 otherwise, $\mathbf{1}_i$ is a unit vector pointing in the direction i and

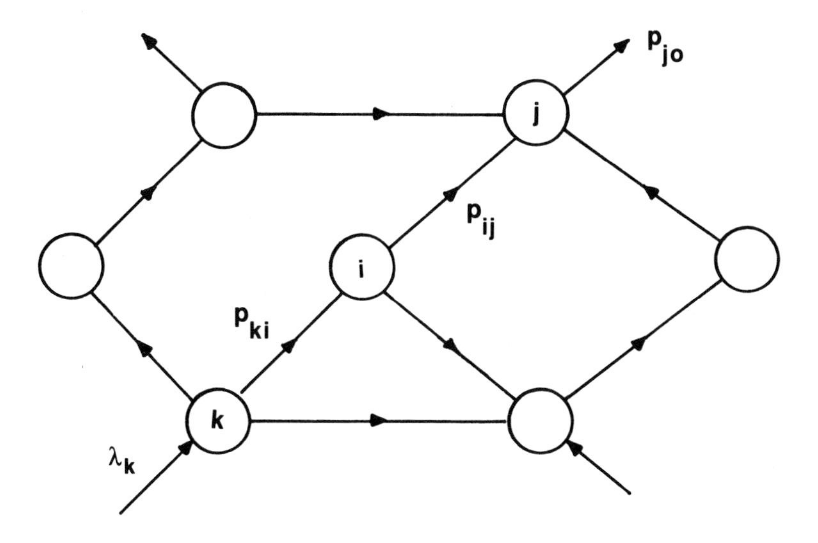

Figure 2.1
The Jackson Queueing Network

$$\beta_i(a) = \begin{cases} a t_i^{-1}, & \text{if } 0 \le a \le k_i, \\ k_i t_i^{-1}, & \text{if } a > k_i. \end{cases}$$

By verification of the above balance equation (eq. 2.1), Jackson proved that the general solution for $\pi(\mathbf{n})$, when it exists, is

$$\pi(\mathbf{n}) = \prod_{i=1}^{N} f_i(n_i), \tag{2.2}$$

where

$$f_i(n_i) = \begin{cases} C_i(k_i\rho_i)^{n_i}/n_i!, & \text{if } 0 \le n_i \le k_i, \\ C_i k_i^{k_i} \rho_i^{n_i}/k_i!, & \text{if } n_i \ge k_i+1, \end{cases}$$

C_i is a normalization constant chosen so that $\sum_{a=0}^{\infty} f_i(a) = 1$, $\rho_i = T_i t_i / k_i$ and T_i is the total throughput of customers at service center i, as given by the solution to the set of traffic flow equations

$$\{T_i = \lambda_i + \sum_{j=1}^{N} p_{ji} T_j \mid i = 1,...,N\}.$$

For the equilibrium distribution $\pi(\mathbf{n})$ to exist, it is required that C_i be nonzero. This is ensured if $\sum_{a=k_i+1}^{\infty} \rho_i^a$ is finite, or equivalently if $\rho_i < 1$.

As can be seen from the form of eq. 2.2, the queue-lengths at the nodes in a Jackson network are independent random variables. Furthermore, each queue behaves as if it were an independent FCFS exponential multi-server queue with Poisson arrivals at rate T_i since the state-distribution for such an isolated queue is given precisely by $f_i(n_i)$. It may also be seen from eq. 2.2 that the state-distribution only depends on the routing probabilities p_{ij} and external arrival rates λ_i through the quantities T_i.

In a later work [JAC2], Jackson extended the original open network of exponential multi-server queues to include servers with state-dependent service-rates, input arrival rates that may depend on the total population of the network, a certain form of state-dependent routing (or blocking) in which a customer may bypass a node if the queue-length has reached a certain limit and, finally, '*triggered arrivals*' where a customer is injected automatically into the network when the departure of a customer from the network has caused the global network population to fall below some fixed threshold.

In the extended open Jackson network the queues at the nodes are assumed to be FCFS, as in the original Jackson network, but it is assumed that there is a single-server. The service-rate μ_i of the server at node i is allowed to take on the state-dependent form $\mu_i(n_i)$, where n_i is the total number of customers at node i, that is, when the population of node i is n_i, the rate at which work is accomplished is $\mu_i(n_i)$. When $\mu_i(n_i) = \text{Min}\{n_i,k_i\}$, we have the case of a FCFS queue with k_i parallel servers, as in the original Jackson network. The input rate of external arrivals to node i is allowed to take on the state-dependent form $\lambda_i(K)$, where $\lambda_i(K) = \gamma(K)(r_i/\sum_{j=1}^{N} r_j)$, $\{r_i \mid i = 1,...,N\}$ is some set of nonnegative constants, $\gamma(K)$ is some function such that $\gamma(K) \geq 0$ for $K \geq 0$ and K is the total population of customers in the network, that is, $K = (n_1+...+n_N)$. The interpretation is that $\gamma(K)$ is the total external arrival rate of customers when the network population is K and $r_i / \sum_{j=1}^{N} r_j$ is the proportion that is directed towards node i. Furthermore, in the particular form of state-dependent routing (or blocking) that is included, a customer bypasses node i, as if he had been serviced, and proceeds on according to the routing probabilities p_{ij}, $j \in \{0,1,...,N\}$, as in the original Jackson network, if upon arrival at node i it is found that $n_i =$

κ_i, where κ_i is some nonnegative integer corresponding to the total number of customers that may ever be present at node i. Finally, in the extended Jackson network, there is a so-called triggered arrivals mechanism whereby a customer is injected automatically into the network if the departure of a customer from the network has caused the network population K to fall below some threshold K^*. The triggered arrival is injected into node i with probability $r_i / \sum_{j=1}^{N} r_j$. In effect, during the time that $K = K^*$, the network behaves as a closed one with routing probabilities p_{ij}^*, given by

$$p_{ij}^* = p_{ij} + p_{i0} r_j \Big/ \sum_{k=1}^{N} r_k,$$

where $i,j \in \{1,\ldots,N\}$, as illustrated in Figure 2.2. In the above generalizations, it is assumed implicitly that $K^* \leq (\kappa_1+\ldots+\kappa_N)$.

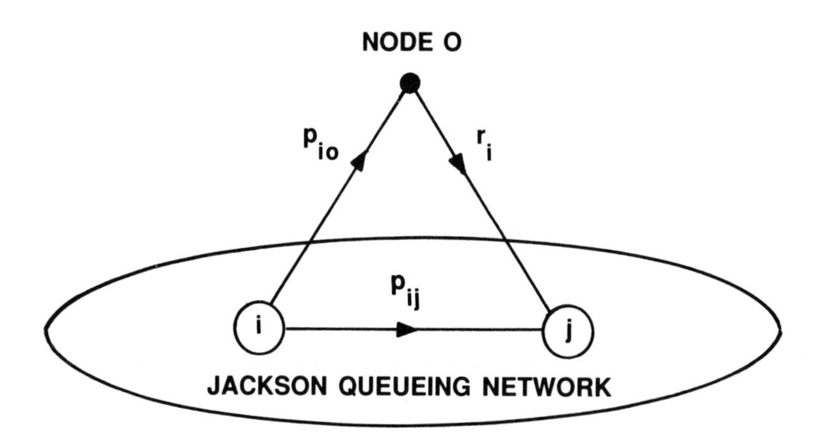

Figure 2.2
Injection Probabilities with Triggered Arrivals

Furthermore, it follows that $\gamma(K)$ need only be defined for $K^* \le K < K_0$, where $K_0 = \text{Min}\{k \mid \gamma(k) = 0, K^* \le K_0 \le (\kappa_1+...+\kappa_N)\}$.

For the extended Jackson network, one may write equilibrium balance equations for the feasible state probabilities, as has been done above for the original Jackson network. The state-space S_e of the extended Jackson network is

$$S_e = \{\mathbf{n} \mid \mathbf{n} = (n_1,...,n_N), 0 \le n_i \le \kappa_i, i = 1,...,N, K^* \le (n_1+...+n_N) \le K_0\}.$$

By verification of the balance equations, it may be proved (see [DIJ1]) that the equilibrium state-distribution $\pi_e(\mathbf{n})$, when it exists, is given by

$$\pi_e(\mathbf{n}) = C^{-1} \{ \prod_{k=K^*}^{(n_1+...+n_N-1)} \gamma(k) \} \prod_{i=1}^{N} g_i(n_i), \tag{2.3}$$

where $\mathbf{n} \in S_e$, $g_i(n_i) = (e_i t_i)^{n_i} / \prod_{a=1}^{n_i} \mu_i(a)$, t_i is the mean service-requirement of the customers at node i, C is a normalization constant and $\{e_i \mid i = 1,...,N\}$ is a set of quantities that satisfy the set of equations

$$\{e_i = (r_i / \sum_{j=1}^{N} r_j) + \sum_{j=1}^{N} e_j p_{ji} \mid i = 1,...,N\} \tag{2.4}$$

(Here, and in the following, an empty product is assumed to take on the value one). As a result, the state-distribution only depends on the routing topology of the network through the quantities e_i.

The normalization constant C in eq. 2.3 is, by definition,

$$C = \sum_{\mathbf{n} \in S_e} \{ \prod_{k=K^*}^{(n_1+...+n_N-1)} \gamma(k) \} \prod_{i=1}^{N} g_i(n_i)$$

or

$$C = \sum_{k=K^*}^{K_{max}} \{ \prod_{a=K^*}^{k-1} \gamma(a) \} C(k),$$

where $K_{max} = \text{Min}\{(\kappa_1+...+\kappa_N), K_0\}$,

$$C(k) = \sum_{\mathbf{n} \in S(k)} \prod_{i=1}^{N} g_i(n_i)$$

and

$$S(k) = \{\mathbf{n} \mid \mathbf{n} = (n_1,...,n_N), 0 \le n_i \le \kappa_i, i = 1,...,N, \sum_{j=1}^{N} n_j = k\}.$$

From eq. 2.3, it follows that the marginal probability $\Pi(x)$ that the network population is x, where $K^* \le x \le K_{max}$, is

$$\Pi(x) = C^{-1} \{ \prod_{k=K^*}^{x-1} \gamma(k) \} C(x).$$

Hence, we may write $\pi_e(\mathbf{n})$ in the simpler form

$$\pi_e(\mathbf{n}) = \Pi(n_1+...+n_N)C(n_1+...+n_N)^{-1} \prod_{i=1}^{N} g_i(n_i). \tag{2.5}$$

A sufficient condition for the equilibrium to exist is that $C < \infty$. For general values of κ_i, $\mu_i(n_i)$ and $\lambda_i(K)$, however, the determination of a simple closed form for C may be difficult.

The extended Jackson network reduces to the original Jackson network when

$$r_i = \lambda_i / \sum_{j=1}^{N} \lambda_j, \quad \gamma(K) = \sum_{j=1}^{N} \lambda_j, \quad K^* = 0, \quad K_0 = \infty, \quad \kappa_i = \infty$$

and $\mu_i(n_i) = \text{Min}\{n_i,k_i\}$, where k_i is the number of parallel servers at node i and t_i is the mean service-requirement of customers at node i.

Another special case of particular interest is when $K_0 = K^*$. In this situation, $\lambda_i(K_0) = 0$. In this network, the equilibrium population is K^* and there are no exogeneous arrivals. For any initial condition, such that $(n_1+...+n_N) \geq K^*$, the population will eventually reach K^* due to departures from the network, resulting in a closed queueing network with a finite fixed population K^*. Once the population of the closed network has reached K^*, a triggered arrival occurs whenever there is a departure from the network. This cancels the effect of any departure and maintains the network population at K^*. In the following section, we consider this particular closed queueing network in more detail.

2.1.2 The Gordon-Newell Queueing Network

Subsequent to the work of Jackson, the closed Jackson queueing network was discovered independently by Gordon and Newell [GOR1], who were apparently unaware of the earlier work of Jackson [GOR2]. In the formulation of Gordon and Newell, however, the state-dependent routing mechanism that had been included by Jackson was not considered. Today, the special case of the closed Jackson queueing network, with no state-dependent routing, is commonly known as the Gordon-Newell network.

The equilibrium distribution $\pi_c(\mathbf{n},K^*)$ of the Gordon-Newell closed queueing network with population K^* and state-space $S(K^*)$, may be derived conveniently from eq. 2.3 by conditioning $\pi_e(\mathbf{n})$ on the event that $\mathbf{n} \in S(K^*)$, where $S(k)$ is as defined in Subsection 2.1.1. From eq. 2.3, we may write

$$\pi_c(\mathbf{n},K^*) = \pi_c(\mathbf{n} \mid n_1+...+n_N = K^*) = C(K^*)^{-1} \prod_{i=1}^{N} g_i(n_i), \tag{2.6}$$

where

$$C(K^*) = \sum_{\mathbf{n} \in S(K^*)} \prod_{i=1}^{N} g_i(n_i).$$

Since the cardinality of the state-space of the Gordon-Newell network is denumerable and finite, $C(K^*)$ is always finite and the state-distribution, as given by eq. 2.6, always exists.

Looking back now at eq. 2.5, we see that eq. 2.3 may be written in the simpler form

$$\pi_e(\mathbf{n}) = \Pi(n_1+...+n_N)\pi_c(\mathbf{n},n_1+...+n_N).$$

Hence, by simply deconditioning the state-distribution of the Gordon-Newell network, we obtain the state-distribution for the extended Jackson network.

If one is given a set of routing probabilities $\{p_{ij}^* \mid i,j = 1,...,N\}$ for a closed network with N nodes, then to find the e_i, appearing in $g_i(.)$, and conform with the formulation of the previous section, we may let $p_{ij} = p_{ij}^*$ for $i = 1,...,N\text{-}1$, $j = 1,...,N$, $p_{i0} = 0$ for $i \neq N$, $p_{Ni} = 0$ for $i = 1,...,N$, $p_{N0} = 1$ and $r_i = p_{Ni}^*$. Using these values for the p_{ij}, we may find e_i, $i = 1,...,N$, using eq. 2.4. Equivalently, the e_i are given by a nontrivial solution to the set of equations

$$\{e_i = \sum_{j=1}^{N} e_j p_{ji}^* \mid i = 1,...,N\}. \tag{2.7}$$

Here the system of linear equations is underdetermined and the e_i are only unique up to a multiplicative constant. However, as can be seen from eq. 2.6, the state-distribution itself only depends on the relative values of the e_i since any multiplicative constant may be absorbed into $C(K^*)^{-1}$. The method of determining the e_i using eq. 2.4, as has been described above, is equivalent to finding the e_i using eq. 2.7 under the constraint that $e_N = 1$. Indeed, if we assume that $e_N = 1$, then we may obtain eq. 2.4 from eq. 2.7 as follows:

$$e_i = \sum_{j=1}^{N} e_j p_{ji}^* = e_N p_{Ni}^* + \sum_{j=1}^{N-1} e_j p_{ji}^* = r_i + \sum_{j=1}^{N} e_j p_{ji}, \tag{2.8}$$

since, by assumption, $r_i = p_{Ni}^*$, $e_N = 1$ and $p_{Ni} = 0$.

From eq. 2.7, we see that the quantity e_i may be interpreted as the relative frequency, or '*visit-ratio*,' with which the customers in a closed network visit node i. Hence, if T_i is the actual throughput of the customers at node i, then $T_i = e_i T_j / e_j$.

In the special case that $\mu_i(n_i) = 1$, a simple closed form is known for the normalization constant $C(K^*)$. In an early paper, Koenigsberg [KOE1] showed that, for a cyclic network of single-server FCFS exponential queues

$$C(K^*) = \sum_{i=1}^{N} \{ X_i^{K^*+N-1} / \prod_{\substack{j=1 \\ j \neq i}}^{N} (X_i - X_j) \}, \tag{2.9}$$

where $X_i = t_i / t_1$. The cyclic network is, however, simply a special case of a Gordon-Newell network with $e_i = 1$ for $i = 1,...,N$. Furthermore, from eq. 2.6, we see that the state-distribution $\pi_c(\mathbf{n}, K^*)$ for a Gordon-Newell network with $\mu_i(n_i) = 1$ depends only on the quantities e_i and t_i through the product $e_i t_i$ since, in this situation,

$$g_i(n_i) = (e_i t_i)^{n_i}.$$

As a result, we may convert any Gordon-Newell network, with visit-ratios e_i and mean service-time requirements t_i, to an equivalent cyclic network, as far as the state-distribution is concerned, by means of the simple transformations:

$$t_i \leftarrow e_i t_i$$
$$e_i \leftarrow 1.$$

It now follows immediately that the normalization constant for a Gordon-Newell network with constant service-rates and population K^* is also given by eq. 2.9, but with $X_i = e_i t_i / e_1 t_1$. This result, which is apparently not well-known [KOE2], may be attributed to Harrison [HAR1], Moore [MOO1] and Swersey [SWE1].

2.1.3 The BCMP Queueing Network

The class of product-form queueing networks was extended greatly and presented in a unified fashion by Baskett, Chandy, Muntz and Palacios [BAS1] in 1975. This class of queueing networks, now commonly known as the class of BCMP queueing networks, has been the one most widely referred to in practice and includes, as special cases, the previous developments that had been made in the area of queueing networks such as the extended Jackson network (with the exception of the special type of state-dependent routing described in Subsection 2.1.1). The paper of Baskett et al is, perhaps, the most cited paper in the field and may be viewed as the main archival reference for equilibrium results on product-form queueing networks. It is to be noted, however, that the principal results of Baskett et al, on the state-distribution, are also contained in a paper by Reiser and Kobayashi [REI3]. Furthermore, at about the same time these papers appeared, Kelly [KEL1,KEL2] published essentially equivalent results that differ mainly in the allowed service disciplines and service-requirement distributions at the nodes. The papers of Kelly, however, give a more theoretical exposition while those of Baskett et al, and Reiser and Kobayashi, provide more explicit results on state-distributions of practical interest. In the following, we shall summarize the modeling features and state-distribution of the class of BCMP queueing networks. The main additional features that may be accommodated, given the results of Kelly, are considered in the following subsection.

In the class of BCMP queueing networks, one may have different types of customers that may be distinguished by their routing parameters, service-time requirements and the possible class memberships that may be taken on. It is assumed that there are R different *types* of customers. A customer of type r may, in time, evolve through different class memberships within a finite set of classes C_r. The sets of classes $C_1,...,C_R$ are assumed to be mutually disjoint. The class membership of a customer is used to index the various possibly different service-time requirements and routing parameters that a customer may have at a node. A customer of type r and in class c at node i which completes its service-requirement proceeds next to node j in class d with probability $p_{ic;jd}^{(r)}$ or leaves the network with probability $p_{ic;0}^{(r)}$, where

$$p_{ic;0}^{(r)} = (1 - \sum_{j=1}^{N} \sum_{d \in C_{jr}} p_{ic;jd}^{(r)}),$$

N is the number of nodes in the network and C_{jr} is the set of classes that a customer of type r may belong to when at node j. Clearly, $C_r = C_{1r} \cup ... \cup C_{Nr}$. The mean service-requirement for a customer of type r in class c at node i is denoted by $m_{ic}^{(r)}$. The transition probability matrix $\mathbf{P}^{(r)} = [p_{ic;jd}^{(r)} : 1 \leq i,j \leq N, c \in C_{ir}, d \in C_{jr}]$ is commonly known as the *routing chain* for type r customers. Customers of type r are commonly referred to as '*chain r*' customers. A BCMP network with multiple types of customers is commonly termed a *multiple-chain* queueing network.

In a BCMP queueing network, we may also have customers which arrive from outside the network according to a Poisson process. We denote the rate of arrival to node i of type r customers in class c by $\lambda_{ic}^{(r)}$. If $\lambda_{ic}^{(r)} > 0$ for some $c \in C_{ir}$ and $i \in \{1,...,N\}$, then we say that the routing chain r is *open*. If $\lambda_{ic}^{(r)} = 0$ for all $c \in C_{ir}$ and $i = 1,...,N$, then we say that routing chain r is *closed*. In this case, it is assumed implicitly that $\mathbf{P}^{(r)}$ defines an irreducible, discrete-time Markov chain with state-space $\{(i,c) \mid c \in C_{ir}, i = 1,...,N\}$. If, in a network, some routing chains are open and others are closed, then we say that the network is *mixed*. If routing chain r is closed, then there is a certain fixed population K_r of type r customers that circulate indefinitely among the nodes of the network and we say that K_r is the population of chain r.

Using the class-switching feature of BCMP queueing networks, it is possible to realize a wide variety of different routing schemes in a network including random routing, deterministic routing and routing in which the future route depends on the past history.

In addition to the features described above, there are a number of different queueing disciplines that may be allowed at the nodes in a BCMP network, namely, first-come first-served (FCFS), processor-sharing (PS), last-come first-served preemptive resume (LCFSPR), where service is resumed at the point of interruption, and infinite-server (IS). At an IS queue there is an unlimited number of servers. If node i has the FCFS queueing discipline, then it is required that all of the service-requirement distributions of the customers at the node be exponential and that $m_{ic}^{(r)}$ be the same for all $r = 1,...,R$ and $c \in C_{ir}$. For the other service disciplines, the service-requirement distribution may take on a very general form as is to be described below.

The required form for the service-requirement distribution at nodes with a PS, LCFSPR or IS queueing discipline is that it have a rational Laplace transform. This includes distributions of the exponential, hyperexponential and hypoexponential type. Cox [COX1] has shown that any distribution with a rational Laplace transform may be constructed using a sequence of exponential stages, as illustrated schematically in Figure 2.3. In this representation, a customer commences service in the first stage. The mean service-requirement at node i for a type r customer in class c at stage s is denoted by $\tau_{ics}^{(r)}$. When this stage is completed, the customer proceeds on to stage (s+1) with probability $\sigma_{ics}^{(r)}$ or completes its overall service-requirement with probability $(1-\sigma_{ics}^{(r)})$.

The general service-requirement distribution described above, which is applicable at queues of the PS, LCFSPR and IS types, may also be constructed by simply assuming exponential service-time requirements and using the class-switching feature of BCMP queueing networks. This may be done as follows. Consider a queue i of the PS, LCFSPR or IS type. Whenever a customer *first* enters such a queue, it immediately begins to receive service. Now suppose that when a type r customer first enters node i in class c it immediately changes to the class index (c,1) with a service-time requirement $\tau_{ic1}^{(r)}$. Once this requirement is met, the customer either changes to class (c,2) with probability $\sigma_{ic1}^{(r)}$ and immediatly returns to node i for another round of service with mean requirement $\tau_{ic2}^{(r)}$ or it leaves the node with probability $(1-\sigma_{ic1}^{(r)})$. In general, we assume that when a type r customer of class (c,s) completes its service-requirement $\tau_{ics}^{(r)}$, it either changes to class (c,s+1) with probability $\sigma_{ics}^{(r)}$ for another round of service at node i with mean requirement $\tau_{ic(s+1)}^{(r)}$ or it leaves node i with probability $(1-\sigma_{ics}^{(r)})$. This mechanism, however, is seen to be equivalent entirely to having

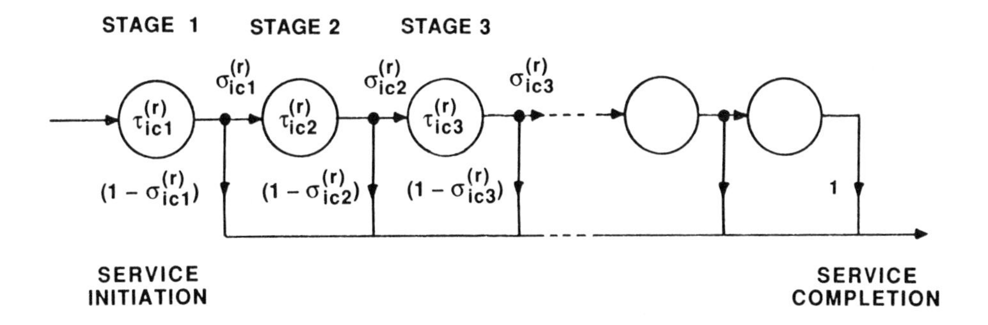

Figure 2.3
Coxian Representation of a Service-Requirement Distribution

a Coxian service-requirement distribution. Hence, by augmenting the class index of a customer with the supplementary variable s and using the class-switching feature, we can synthesize general service-requirement distributions of the Coxian type. The possibility of accommodating general service-requirement distributions of this type at PS, LCFSPR and IS nodes may, therefore, be viewed simply as being a consequence of the class-switching feature.

The state description of a BCMP network includes the number of each class of customers at the nodes, the stage of service they have reached and the ordering of the customers at queues of the FCFS and LCFSPR type. Baskett et al [BAS1] give the equilibrium state-distribution for the network at this level of detail. They also provide the marginal distribution for the joint number of each class of customers at the nodes. For networks in which all routing chains are open and in which the rates of external arrivals are constant, they give expressions for the marginal distribution of the joint total number of customers at the nodes. In the following, we shall only consider the marginal distribution of the joint number of customers of each *type* at the nodes for closed and mixed networks since this is the distribution of most interest in practice.

Let n_{ir} be the total number of type r customers at node i, irrespective of their actual class membership, and let $n_i^{(R)} = \sum_{r=1}^{R} n_{ir}$, $n^{(r)} = \sum_{i=1}^{N} n_{ir}$, $\mathbf{n}_i^{(R)} = (n_{i1},\ldots,n_{iR})$ and $\mathbf{n}^{(R)} = (\mathbf{n}_1^{(R)},\ldots,\mathbf{n}_N^{(R)})$. We shall also sometimes abbreviate $n_i^{(R)}$ as n_i. The marginal state-distribution, when it exists, of the joint number of customers of each type at the nodes in a BCMP queueing network, as has been described above, is given by

$$\pi(\mathbf{n}^{(R)}) = G_1^{-1} \prod_{i=1}^{N} f_i(\mathbf{n}_i^{(r)}), \tag{2.10}$$

where $\mathbf{n}^{(R)} \in S^{(R)}$, $S^{(R)}$ is the state-space of the queueing network and G_1 is a normalization constant. Furthermore, in the above

$$f_i(\mathbf{n}_i^{(r)}) = \begin{cases} n_i^{(R)}! \prod_{r=1}^{R} w_{ir}^{n_{ir}}/n_{ir}!, & \text{if node i is FCFS, PS or LCFSPR,} \\ \prod_{r=1}^{R} w_{ir}^{n_{ir}}/n_{ir}!, & \text{if node i is IS,} \end{cases}$$

$$w_{ir} = e_{ir}t_{ir}, \qquad e_{ir} = \sum_{c \in C_{ir}} \alpha_{ic}^{(r)}, \qquad t_{ir} = \sum_{c \in C_{ir}} \alpha_{ic}^{(r)} m_{ic}^{(r)}/e_{ir},$$

where the quantities $\alpha_{ic}^{(r)}$, $i = 1,...,N$, $c \in C_{ir}$, satisfy the set of equations

$$\{\alpha_{ic}^{(r)} = \lambda_{ic}^{(r)} + \sum_{j=1}^{N} \sum_{d \in C_{jr}} \alpha_{jd}^{(r)} p_{jd;ic}^{(r)} \mid c \in C_{ir}, i = 1,...,N\} \tag{2.11}$$

and the state-space $S^{(R)}$ is

$$S^{(R)} = \{\mathbf{n}^{(R)} \mid n_{ir} \geq 0, n^{(r)} = K_r \text{ if routing chain r is closed, } i = 1,...,N, \; r = 1,...,R\}.$$

The quantity w_{ir} is commonly referred to as the *relative traffic intensity* of type r customers at node i. The normalization constant G_1 is, by definition,

$$G_1 = \sum_{\mathbf{n}^{(R)} \in S^{(R)}} \prod_{i=1}^{N} f_i(\mathbf{n}_i^{(r)}).$$

If routing chain r is open, then the quantity $\alpha_{ic}^{(r)}$, appearing in eq. 2.11, may be interpreted as the actual throughput of type r customers of class c at node i, assuming that $\pi(\mathbf{n}^{(R)})$ exists. If, however, routing chain r is closed, then $\alpha_{ic}^{(r)}$ is unique only up to a multiplicative constant and, in this situation, it may be interpreted as a *relative throughput* or 'visit-ratio'.

The distribution, as given by eq. 2.10, always exists in the case for which the network is closed. If the network is open or mixed, then a sufficient condition for the existence of a stationary distribution is that $G_1 < \infty$.

As can be seen from eq. 2.10, the marginal distribution only depends on the service-requirement distributions through their respective means. This is known as the *insensitivity* property. Furthermore, we see that the distribution depends only on $\mathbf{P}^{(r)}$, $m_{ic}(r)$ and $\lambda_{ic}(r)$ through the quantities e_{ir} and t_{ir}. The quantity e_{ir} may be interpreted as the total throughput of type r customers at node i, if routing chain r is open, or as the overall visit-ratio of type r customers at node i, if routing chain r is closed. The quantity t_{ir} is the overall mean service-requirement for type r customers that visit node i. Hence, we see that, as far as the marginal equilibrium state-distribution is concerned, a multiple-chain BCMP queueing network with class-switching and Coxian service-requirement distributions at IS, PS and LCFSPR nodes may be mapped, almost trivially, into an equivalent network having exponential distributions and no class-switching [REI3].

In addition to the modeling features described above, BCMP queueing networks may also accommodate several forms of state-dependency, as in the extended Jackson network. The rates of external Poisson arrivals may be made to depend on the population of the network and the rate at which work is accomplished at a node may be made to depend on the number of each type of customers at the node. In addition, the total rate at which work is accomplished in a subset of nodes may be made to depend on the population of that particular subnetwork.

If $\lambda(x)$ is the *total* external arrival rate of customers to the network when $(n_1+...+n_N) = x$ and $p_{0;ic}{}^{(r)}$ is the proportion directed into node i in class c, where $c \in C_{ir}$, then the equilibrium state-distribution is [BAS1]

$$\pi(\mathbf{n}^{(R)}) = G_2^{-1} d(\mathbf{n}^{(R)}) \prod_{i=1}^{N} f_i(\mathbf{n}_i^{(r)}), \qquad (2.12)$$

where

$$d(\mathbf{n}^{(R)}) = \prod_{a=0}^{(n_1+\ldots+n_N-1)} \lambda(a), \qquad (2.13)$$

$\mathbf{n}^{(R)} \in S^{(R)}$, G_2 is a normalization constant and the quantities $\alpha_{ic}^{(r)}$ are as defined by eq. 2.11, but with $\lambda_{ic}^{(r)}$ replaced by $p_{0;ic}^{(r)}$. Here it is assumed that

$$\sum_{r=1}^{R} \sum_{i=1}^{N} \sum_{c \in C_{ir}} p_{0;ic}^{(r)} = 1.$$

In a BCMP queueing network, we may also have R external Poisson sources that feed the R closed routing chains individually. If $\lambda_r(x)$ is the total arrival rate of exogeneous type r customers when $n^{(r)} = x$ and $q_{0;ic}^{(r)}$ is the proportion directed into node i in class c, where $c \in C_{ir}$, then the state-distribution is the same as eq. 2.12 but with $d(\mathbf{n}^{(R)})$ (see eq. 2.13) given by [BAS1]

$$d(\mathbf{n}^{(R)}) = \prod_{r=1}^{R} \prod_{a=0}^{n^{(r)}-1} \lambda_r(a) \qquad (2.14)$$

and with the quantities $\alpha_{ic}^{(r)}$ defined by eq. 2.11, but with $\lambda_{ic}^{(r)}$ replaced by $q_{0;ic}^{(r)}$. Here it is assumed that

$$\sum_{i=1}^{N} \sum_{c \in C_{ir}} q_{0;ic}^{(r)} = 1.$$

As in the extended Jackson network, we may also have state-dependent service-rates. Consider a node i of the FCFS, PS or LCFSPR type. If $\mu_i(n_i)$ is the rate at which work is accomplished at such a node when the population of the node is n_i, then the state-distribution is the same as eq. 2.10 but with $f_i(\mathbf{n}_i^{(r)})$ replaced by [BAS1]

$$f_i(\mathbf{n}_i^{(r)}) \; / \prod_{a=1}^{n_i} \mu_i(a).$$

If $\mu_{ir}(n_{ir})$ is the rate at which work is accomplished for customers of type r when the number of type r customers at node i is n_{ir}, then we need only replace the above terms by [BAS1]

$$f_i(\mathbf{n}_i^{(r)}) \; / \prod_{r=1}^{R} \prod_{a=1}^{n_{ir}} \mu_{ir}(a).$$

This last type of state-dependency, however, is not allowed at queues of the FCFS type.

If I is some subset of the set of nodes $\{1,\ldots,N\}$, $n_I = \sum_{i \in I} n_i^{(R)}$ and $Z_I(x)$ is a multiplicative factor which modifies the rates at which work is normally accomplished for the customers being serviced in the subnetwork I when $n_I = x$, then the state-distribution is again given by eq. 2.10 but with $\prod_{i \in I} f_i(\mathbf{n}_i^{(r)})$ replaced by [BAS1]

$$\prod_{i \in I} f_i(\mathbf{n}_i^{(r)}) \; / \prod_{a=1}^{n_I} Z_I(a).$$

It is to be noted that it is possible to combine all of the various forms of state-dependency mentioned above and maintain a product-form state-distribution [BAS1]. Finally, it is to be mentioned that in the class of BCMP queueing networks we may not accommodate service-rate functions $\mu_i(.)$ that depend in general on $\mathbf{n}_i^{(R)}$. There exist, however, functions of the general form $\mu_{ir}(\mathbf{n}_i^{(R)})$ with a pathological structure that do support a product-form state-distribution [BAL1,NEU1].

2.1.4 Generalizations of Kelly

Kelly [KEL1,KEL2] has formulated a class of queueing networks that is equivalent essentially to that of Baskett et al, apart from the allowed types of queueing disciplines and the allowed class of service-requirement distributions. The marginal state-distributions in both classes of networks are, however, the same. In the following, we shall only describe the main additional features offered by Kelly that are not included in the class of networks considered by Baskett et al.

In the queueing network of Kelly, the queueing disciplines at the nodes may take on a very general form that includes the FCFS, PS, LCFSPR and IS disciplines as special cases. In the general formulation of Kelly, it is assumed that the customers at node i are ordered from 1 to n_i, where n_i is the total number of customers at node i. A customer who arrives at node i moves into position p with probability $\delta_i(p,n_i+1)$ and the customers already in the queue, who were previously in the positions $p,...,n_i$, move back to positions $p+1,...,n_i+1$, respectively. The rate at which work is accomplished at the node, when there are n_i customers present, is $\mu_i(n_i)$ and the proportion of this work rate given to the customer in position p is $\gamma_i(p,n_i)$. For simplicity, we assume that $\mu_i(n_i) > 0$ for $n_i > 0$. It is, however, feasible to include the possibility that $\mu_i(n_i) = 0$ for certain values of n_i [KEL1]. In the general formulation of Kelly, it is also assumed that $m_{ic}^{(r)} = 1$ for all $c \in \mathcal{C}_{ir}$ and $r = 1,...,R$ and that the service-requirement distributions of all customers are exponential. The condition that $m_{ic}^{(r)} = 1$ may seem restrictive. However, if $m_{ic}^{(r)} = t$, where t is some positive constant that is independent of c and r, then we may replace t by 1, $\mu_i(n_i)$ by $t^{-1}\mu_i(n_i)$ and achieve the same effect as if $m_{ic}^{(r)}$ were equal to t.

By suitable choice of the functions $\delta_i(.)$, $\gamma_i(.)$ and $\mu_i(.)$, we may construct various types of queueing disciplines such as FCFS, PS, LCFSPR, IS and random order service (see [KEL3] for examples). By placing certain restrictions on these functions, more general service-requirement distributions may be accommodated.

If $\delta_i(p,n_i+1) = \gamma_i(p,n_i+1)$, then we have a so-called *symmetric* queue and we may allow the service-requirement distribution of a customer to take on a general form that may be dependent on r and c. Examples of symmetric queues include PS, LCFSPR and IS queues. The allowed generalized distribution is assumed to be made up of $z_{ic}^{(r)}$ phases of service-requirements, each of which are exponentially distributed with mean $\zeta_{ic}^{(r)}$, respectively, such that $z_{ic}^{(r)}\zeta_{ic}^{(r)} = m_{ic}^{(r)}$. This corresponds to a Gamma distribution [COO1] with mean $m_{ic}^{(r)}$. It is apparent, however, that by making use of the class-switching feature of BCMP networks, we may also synthesize service-requirement distributions that are made up of a finite mixture of gamma distributions. This may be done in the same way as was done in Subsection 2.1.3 where we constructed the Coxian distribution using the class-switching feature. Since a general distribution may be approximated by a finite mixture of gamma distributions, with a mean-square error that may be made arbitrarily small, Kelly conjectured that the service-time requirement distributions may, in fact, be arbitrarily distributed at symmetric queues. This conjecture was proved by Barbour [BAR1].

The marginal state-distribution of the joint total number of each type of customers at the queues, when the queueing discipline is of the general type described above, is identical in form to that of eq. 2.10 with $f_i(\mathbf{n}_i^{(r)})$ corresponding to the case of a PS queue having a state-dependent service-rate function $\mu_i(n_i)$. If there are, in addition, state-dependent arrival rates, then the state-distribution has the same form as eq. 2.12.

Kelly [KEL2] also considers the special type of state-dependent routing (or blocking) mechanism that had been introduced by Jackson.

2.1.5 Generalized Population Size Constraints

Subsequent to the works of Baskett et al, Reiser and Kobayashi, and Kelly, Lam [LAM4] generalized the mechanism of triggered arrivals and the mechanism whereby exogeneous customers may arrive at a BCMP queueing network.

As in the original BCMP network, it is assumed that there is either

(1) a single external source of Poisson arrivals with a rate $\lambda(.)$ that depends on the total population of the network, or

(2) several independent external Poisson sources that feed the open routing chains individually.

In the second case, if routing chain r is open, then it is assumed to be fed with customers of type r and the rate of arrival $\lambda_r(.)$ is allowed to depend on the total population of type r customers in the network. In the first case, a proportion $p_{0;ic}^{(r)}$ of the arrivals is directed to node i in class c, where $c \in C_{ir}$. In this case, we have

$$\sum_{r=1}^{R} \sum_{i=1}^{N} \sum_{c \in C_{ir}} p_{0;ic}^{(r)} = 1.$$

In the second case, a proportion $q_{0;ic}^{(r)}$ of the type r arrivals is directed to node i in class c, where $c \in C_{ir}$. In this case, we have

$$\sum_{i=1}^{N} \sum_{c \in C_{ir}} q_{0;ic}^{(r)} = 1.$$

The first case is seen to be equivalent to a situation in which we have a number of independent Poisson sources, as in the second case, but where λ_r depends on the total population of customers in the network and

$$q_{0;ic}^{(r)} = p_{0;ic}^{(r)} / \sum_{j=1}^{N} \sum_{d \in C_{jr}} p_{0;jd}^{(r)}.$$

In the network of Lam, it is also assumed that there is a mechanism of triggered arrivals, as in the extended Jackson network. The Jackson network, however, only assumes a single type of customer and Baskett et al did not generalize the triggered arrival mechanism of Jackson for multiple open routing chains. Nevertheless, as has been noted by Lam [LAM4], the triggered arrival mechanism can be generalized readily to the case of multiple types of customers. There are several generalizations that may be considered. In one possible situation, if a customer of type r leaves the network when $(n_1+...+n_N) = K^*$, where K^* is some constant, then this type r customer is injected immediately into node i in class c, where $c \in C_{ir}$, with probability $q_{0;ic}^{(r)}$. Using this mechanism, in conjunction with the first type of state-dependent arrival process, we may constrain the total population of a BCMP queueing network to lie within a certain range. If we cause the automatic injection of a type r customer when $n^{(r)} = K_r^*$, where K_r^* is some constant indexed by r, and use the second type of state-dependent arrival process, then we may constrain the populations of the open chains individually. We may also consider the case in which we have the first type of arrival process and where we cause the injection of a type r customer when $n^{(r)} = K_r^*$. In this situation, we have individual lower bounds on the populations of the open chains and a global upper bound on the total population of the network. All of these generalizations for BCMP queueing networks follow readily from the results of Jackson. Lam, however, has formulated a more general form of state-dependency that includes all of the above mentioned generalizations as special cases.

In the formulation of Lam, there are so-called *loss functions* $\mathbf{L}(.)$ and *trigger functions* $\mathbf{T}(.)$ that govern the external arrival processes and the triggered arrivals. If an exogeneous customer of type r arrives at the network and $\mathbf{L}_r(\eta^{(R)}) = 1$, where $\eta^{(R)} = (n^{(1)},...,n^{(R)})$ is the *network population vector*, then the arrival is accepted into the network. If $\mathbf{L}_r(\eta^{(R)}) = 0$, then the arriving customer is blocked and lost. If a customer of type r departs from the network when $\mathbf{T}_r(\eta^{(R)}) = 0$, then the departing customer of type r is injected immediately into node i in class c, $c \in C_{ir}$, with probability $q_{0;ic}{}^{(r)}$. If $\mathbf{T}_r(\eta^{(R)}) = 1$, then there is no triggered arrival and the departing customer leaves the system.

The loss and trigger functions of Lam generalize greatly the manner in which the population vector $\eta^{(R)}$ is allowed to evolve in time. If we assume that the queueing network is in equilibrium and that P is the set of feasible population vectors $\eta^{(R)}$, then in order for a product-form distribution to hold, Lam has proved that it is sufficient that $\mathbf{T}_r(\eta^{(R)}) = 1$ *if and only if* $\mathbf{L}_r(\eta^{(R)}-\mathbf{1}_r) = 1$, where $\eta^{(R)} \in P$, $(\eta^{(R)}-\mathbf{1}_r) \in P$ and $\mathbf{1}_r$ is a unit vector pointing in the direction r. In words, if the population vector is $\eta^{(R)}$ and we allow a departure of type r to leave the network, then we must also allow for the acceptance of a type r arrival when the population vector is $(\eta^{(R)}-\mathbf{1}_r)$. If the population vector is $\eta^{(R)}$ and we block the departure of a type r customer by using the triggered arrival mechanism, then we must also block exogeneous arrivals of type r during the time that the population vector is $(\eta^{(R)}-\mathbf{1}_r)$. Figure 2.4 illustrates a space P and a set of possible transitions between population vectors that satisfy the sufficient condition for a product-form distribution. The essential feature of the sufficient condition is that the states of P be 'doubly-connected' or not connected at all.

The marginal equilibrium distribution for a queueing network with lost and triggered arrivals, of the type described above, has the same form as eq. 2.12 with $d(\mathbf{n}^{(R)})$ given by eq. 2.13, if we have the first type of arrival process, and by eq. 2.14, if we have the second type of arrival process. The state-space of the network with lost and triggered arrivals is, however,

$$S_L{}^{(R)} = \{\mathbf{n}^{(R)} \mid n_{ir} \geq 0,\ n^{(r)} = K_r \text{ if routing chain r is closed, } i = 1,\ldots,N, \quad r = 1,\ldots,R,\ \eta^{(R)} \in P\}$$

so that the normalization constant G_2 associated with the state-distribution is changed.

2.1.6 Generalized State-Dependent Routing

Although the class-switching feature of BCMP queueing networks allows for much generality in a routing scheme, it does not permit the routing parameters to vary as a function of the state $\mathbf{n}^{(R)}$ of the network. Towsley [TOW1] was the first to introduce a general form of state-dependent routing that supports a product-form solution for the network equilibrium state probabilities.

In [TOW1], a closed BCMP queueing network with a single type of customer is assumed (R=1). State-dependent routing results for open and closed BCMP queueing networks with multiple types of customers follow from the developments in [TOW1] and may be found in [TOW2]. Subsequent to the works of Towsley, Krzesinski [KRZ1,KRZ2] published state-dependent routing results for closed BCMP queueing networks with multiple types of customers. Although the expositions of Towsley and Krzesinski differ, the state-dependent routing mechanisms described are essentially equivalent. In the following, we shall describe the routing scheme of Towsley and assume a single type of customer. When there are multiple types of customers, the form of the routing scheme is fundamentally

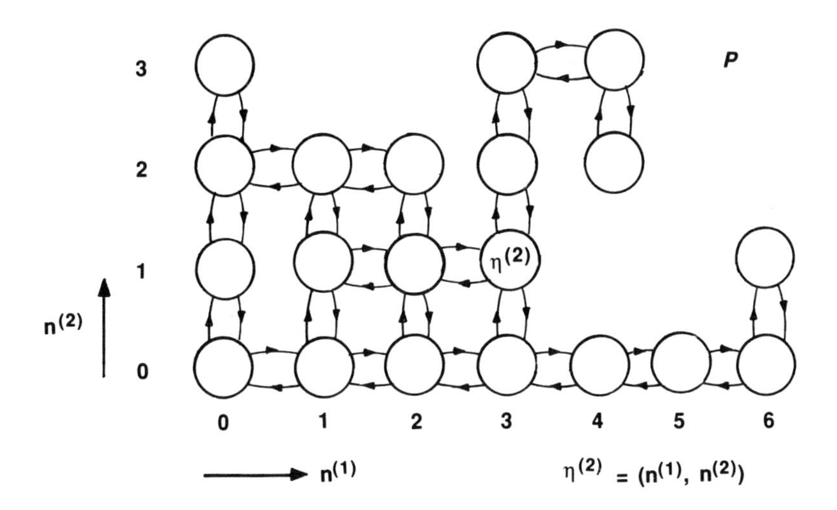

Figure 2.4
Example of a Feasible Population Size Constraint

unchanged except that the routing probabilities for a customer become chain dependent and the various parameters take on an additional chain index r.

In the routing scheme of Towsley for BCMP queueing networks with a single type of customer, it is assumed that there exist so-called parallel subnetworks, or *p-subnetworks*, each having a single entry point and a single exit point. The set of queues contained in p-subnetwork m is denoted by $s(\mathrm{m})$. Each p-subnetwork is assumed to be made up of L_m parallel branches. The set of queues contained within branch b of p-subnetwork m is denoted by $s(\mathrm{m,b})$. It is

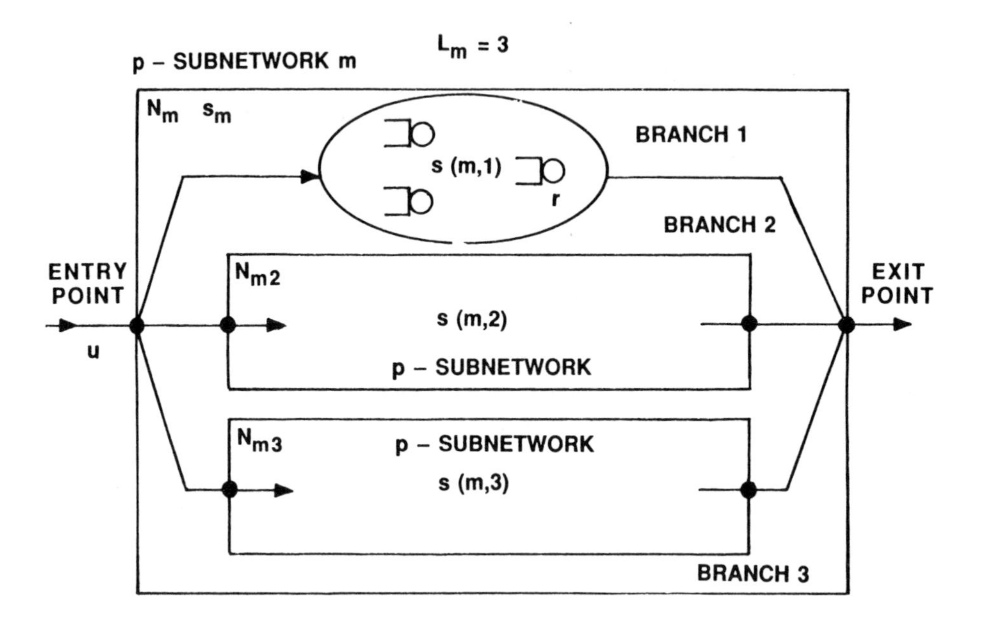

Figure 2.5
Illustration of the Nesting of Parallel Subnetworks

assumed that the sets $s(m,1),\ldots,s(m,L_m)$ are mutually disjoint. Once a customer enters into p-subnetwork m through the particular entry point, it then enters one of the L_m branches and remains in that particular branch until it leaves through the exit point of p-subnetwork m. It is possible for the p-subnetworks to be nested. That is, a particular branch of a p-subnetwork may itself be a p-subnetwork made up of a number of other parallel branches. If m and n are two p-subnetworks, then it is required that either $s(m) \cap s(n) = \emptyset$, $s(m) \supset s(n)$ or $s(n) \supset s(m)$. Figure 2.5 illustrates the nesting of subnetworks.

If u is the entry point for p-subnetwork m and $v \in s(m,b)$, then the transition probability t_{uv} from node u to v may take on the state-dependent form

$$t_{uv} = p_{uv}^* h_m(N_m) h_{mb}(N_{mb}), \tag{2.15}$$

where p_{uv}^* is the nominal transition probability, $h_m(.)$ and $h_{mb}(.)$ are certain non-negative functions to be described, N_m is the total population of p-subnetwork m and N_{mb} is the total population of branch b of p-subnetwork m. If u is the entry point to a set of p-subnetworks P_{uv} to which queue v belongs, that is $v \in s(m)$ for all $m \in P_{uv}$, and b_m is the branch to which v belongs in p-subnetwork m, then the transition probability from node u to v may take on the more general form

$$t_{uv} = p_{uv}^* \prod_{m \in P_{uv}} h_m(N_m) \prod_{m \in P_{uv}} h_{mb_m}(N_{mb_m}). \tag{2.16}$$

In order that the sum of the transition probabilities out of an entry point u sums to one, the functions $h_m(.)$ and $h_{mb}(.)$ are necessarily restricted to the forms

$$h_m(x) = 1 / (xC_m + \sum_{b=1}^{L_m} d_{mb}) \quad \text{and} \quad h_{mb}(x) = xC_m + d_{mb},$$

respectively, where C_m and d_{mb} are arbitrary constants such that $t_{uv} > 0$ when $p_{uv}^* > 0$. The state-dependent routing probability t_{uv} is, therefore allowed to be a rational function of certain state variables. Assuming a single type of customer (R=1), the equilibrium distribution π_T, in the case of state-dependent routing, is

$$\pi_T(\mathbf{n}^{(1)}) = G_3^{-1} \prod_{m=1}^{M} H_m(N_m) \prod_{b=1}^{L_m} H_{mb}(N_{mb}) \prod_{i=1}^{N} f_i(\mathbf{n}_i^{(1)}), \tag{2.17}$$

where the functions $H_m(.)$ and $H_{mb}(.)$ are defined by the recurrence relations

$$H_m(x) = \begin{cases} H_m(x-1)h_m(x-1), \text{ if } x > 0, \\ 1, \text{ if } x = 0, \end{cases}$$

$$H_{mb}(x) = \begin{cases} H_{mb}(x-1)h_{mb}(x-1), \text{ if } x > 0, \\ 1, \text{ if } x = 0. \end{cases}$$

In eq. 2.17, M is the total number of defined p-subnetworks, G_3 is a normalization constant and the quantities e_{i1} appearing in the term $f_i(\mathbf{n}_i^{(1)})$ are a set of (nontrivial) quantities that satisfy the set of equations

$$\{e_{i1} = \sum_{j=1}^{N} e_{j1}\tau_{ji} \mid i = 1,...,N\},$$

where

$$\tau_{ji} = \begin{cases} p_{ji}, & \text{if the transition probability from j to i is constant,} \\ p_{ji}^*, & \text{if the transition probability from j to i is state-dependent.} \end{cases}$$

The specialized forms for the state-dependent transition probabilities, as given by eqs. 2.15 and 2.16, are sufficient for a product-form state-distribution to hold. It is to be noted, however, that other forms of state-dependent routing may exist that also support a product-form distribution.

By suitable choice [TOW1,KRZ1] of the parameters p_{uv}^*, C_m and d_{mb}, a variety of routing mechanisms may be realized including, for example, load balancing (or adaptive routing) and concurrency constraints, in which bounds are placed in a queueing network on the populations of branches and p-subnetworks. Specific examples are illustrated in [TOW1] and [KRZ1].

2.1.7 The MSCCC Queueing Discipline

In a recent paper [LEB1], Le Boudec considered a new type of queueing discipline, of practical importance, that may be included in a BCMP queueing network. In the queue considered by Le Boudec, it is assumed that there are B identical servers that each accomplish work at unit rate and that the service-requirements for all customers that visit the queue are exponentially and identically distributed. When a customer arrives at the queue, it joins one of M groups of classes. An arriving customer of type r joins group m with probability q_{mr} and takes on the class membership (m,r). The service discipline is FCFS, apart from the condition that two customers belonging to the same group may not receive service concurrently. If there is a customer being served whose class membership belongs to group m, then it is said that group m is *active*. When a server becomes available, service is given to the first waiting customer, in first-in first-out (FIFO) order, that belongs to a group which is not active, if any.

The queueing discipline described above has been termed a multiple-server queue with concurrent classes of customers (MSCCC). Such queues are known to support a product-form state-distribution [MARS1,IRA1]. Le Boudec, however, established that the MSCCC queueing discipline could be incorporated within the general framework of the class of BCMP queueing networks.

For a node i which is of the MSCCC type, the state description is specified by $(AC,\mathbf{c})$, where AC is the set of classes (m,r) receiving service, $\mathbf{c} = (c_1,\ldots,c_W)$, c_i is the class of the *i*th customer in the FIFO waiting room and W is the total number of customers waiting for service. An aggregate state description is $\mathbf{x}$, where $\mathbf{x} = [x_{mr} : 1 \leq m \leq$

M, $1 \leq r \leq R$] and x_{mr} is the total number of class (m,r) customers in the node.

For a multiple-chain BCMP queueing network containing MSCCC queues, the marginal state-distribution retains the general form of eq. 2.10 except that if node i is of .the MSCCC type, then $f_i(\mathbf{n}_i^{(R)})$ is replaced by $f_i(\mathbf{x})$, where

$$f_i(\mathbf{x}) = \varphi_B(\mathbf{v}) \prod_{m=1}^{M} v_m! \, (\prod_{r=1}^{R} \rho_{mr}{}^{x_{mr}}/x_{mr}!),$$

$\rho_{mr} = e_{ir} q_{mr} \mu^{-1}$, e_{ir} is the overall visit-ratio for chain r customers at node i, μ^{-1} is the mean service-requirement that is common to all customers that visit the queue, $v_m = \sum_{r=1}^{R} x_{mr}$, $\mathbf{v} = (v_1,...,v_M)$ and $\varphi_B(\mathbf{v})$ is a function defined by the recurrence relation

$$\varphi_B(\mathbf{v}) = \begin{cases} 1, \text{ if } v_m = 0 \text{ for } 1 \leq m \leq M, \\ 0, \text{ if } v_m < 0 \text{ for any } m, \ 1 \leq m \leq M, \\ \sum_{m=1}^{M} \varphi_B(\mathbf{v}-\mathbf{1}_m)/\text{Min}\{nz(\mathbf{v}),B\}, \text{ if } v_m > 0 \text{ for any } m, \end{cases}$$

where $\mathbf{1}_m$ is a unit vector pointing in the direction m and $nz(\mathbf{v})$ is the number of nonzero components in the vector $\mathbf{v}$ [LEB2].

If B = 1, then the MSCCC queue reduces to a single-server FCFS queue. If $B \geq M$, then the MSCCC queue is equivalent to having M parallel FCFS single-server queues with the *m*th queue serving all customers belonging to group m.

2.2 Reversible Queueing Networks

Under certain conditions, to be described below, a Markov chain is said to be *reversible*. If the conditions for reversibility are satisfied in the Markov chain associated with a queueing network, then we have a so-called *reversible queueing network*. A reversible Markov chain has a very simple structure which may be viewed as a multi-dimensional extension of the classical single-dimensional birth-death process. Reversibility is a sufficient condition for a Markov chain to exhibit a product-form equilibrium state-distribution. In this section, we summarize the main results for reversible Markov chains that we shall make use of in subsequent chapters. It is to be noted, however, that our only use of the notion of reversibility will be as a vehicle to simplify our arguments. Indeed, this same vehicle has been used in some early papers (e.g. [KING1]) to facilitate the constructive derivation of the state-distribution of queueing networks.

Consider a stochastic process $X(t)$ with a countable state-space S for $t \in T$. If the distribution of the multi-variate random variable $(X(t_1),...,X(t_n))$ is the same as that of $(X(\tau-t_1),...,X(\tau-t_n))$ for any $t_1,...,t_n$, and $\tau \in T$, then the process $X(t)$ is said to be reversible. If $X(t)$ is reversible, then it is also necessarily stationary since both $(X(t_1),...,X(t_n))$ and $(X(\tau+t_1),...,X(\tau+t_n))$ have the same distribution as $(X(-t_1),...,X(-t_n))$. In [KEL3], it has been proved that a stationary, continuous-time Markov chain with state-space S and infinitesimal generator $\mathbf{Q} = [q_{ij} : i,j \in S]$, is reversible if and only if there exists a set of positive numbers $\{\pi_i \mid i \in S\}$, summing to one, that satisfy the set of equations

$$\{\pi_i q_{ij} = \pi_j q_{ji} \mid i,j \in S\}. \tag{2.18}$$

If a set of numbers $\{\pi_i \mid i \in S\}$ can be found which satisfy eq. 2.18, then $\{\pi_i \mid i \in S\}$ is the equilibrium distribution for the process since eq. 2.18 expresses a *detailed* balance condition which implies that the *global* equilibrium balance equations are also satisfied.

In an analogous manner, if X(t) is a discrete-time Markov chain with state-space S, then it is reversible if and only if there exists a collection of positive numbers $\{\pi_i \mid i \in S\}$, summing to one, that satisfy the set of equations

$$\{\pi_i p_{ij} = \pi_j p_{ji} \mid i,j \in S\},$$

where p_{ij}, $i,j \in S$, is the transition probability from state i to state j. The detailed balance conditions imply that in a reversible Markov chain adjacent states are 'doubly-connected' or not connected at all.

The state-distribution for a reversible, continuous-time Markov chain follows immediately from the relation $\pi_i = \pi_j q_{ji}/q_{ij}$. If 'a', $a \in S$, is an arbitrary reference state and $a, j_1, j_2, \ldots, j_n, i$ is any realizable sequence of states leading from a to i, then

$$\pi_i = C(q_{aj_1} q_{j_1 j_2} \cdots q_{j_n i})/(q_{j_1 a} q_{j_2 j_1} \cdots q_{ij_n}),$$

where C is a normalization constant.

One important property of a reversible Markov chain, and one that we shall make use of, is that if the state-space S is *truncated* to a set A, $S \supset A$, where A is a set of communicating states (that is, if $i \in A$ and $j \in \{S-A\}$, then we change both q_{ij} and q_{ji} to zero), then the resulting Markov chain defined over the state-space A is itself reversible and the equilibrium distribution $\{\pi_i' \mid i \in A\}$, is

$$\pi_i' = \pi_i \Big/ \sum_{j \in A} \pi_j. \qquad (2.19)$$

Under special conditions, a queueing network is reversible. Consider, for the sake of simplicity, a multiple-chain BCMP queueing network with state-independent exogeneous arrivals, exponentially distributed service-time requirements, state-dependent service-rates and no class-switching ($C_{ir} = \{1\}$), and suppose, initially, that all nodes are of the PS type. In this situation, the detailed balance condition between the pair of states $\mathbf{n}^{(R)}$ and $\mathbf{n}_{ij}^*$, where $\mathbf{n}_{ij}^* = (\mathbf{n}_1^{(R)},\ldots,\mathbf{n}_i^{(R)}-\mathbf{1}_r,\ldots,\mathbf{n}_j^{(R)}+\mathbf{1}_r,\ldots,\mathbf{n}_N^{(R)})$, is

$$\pi(\mathbf{n}^{(R)})n_{ir}\mu_{ir}(n_{ir})t_{ir}^{-1}p_{i1;j1}^{(r)}/n_i = \pi(\mathbf{n}_{ij}^*)(n_{jr}+1)\mu_{jr}(n_{jr}+1)t_{jr}^{-1}p_{j1;i1}^{(r)}/(n_j+1), \tag{2.20}$$

where $\mathbf{n}^{(R)} \in S^{(R)}$ and $\mathbf{n}_{ij}^* \in S^{(R)}$. The detailed balance condition between the pair of states $\mathbf{n}^{(R)}$ and $\mathbf{n}_{ir}^-$, where $\mathbf{n}_{ir}^- = (\mathbf{n}_1^{(R)},\ldots,\mathbf{n}_i^{(R)}-\mathbf{1}_r,\ldots,\mathbf{n}_j^{(R)},\ldots,\mathbf{n}_N^{(R)})$, is

$$\pi(\mathbf{n}^{(R)})n_{ir}\mu_{ir}(n_{ir})t_{ir}^{-1}(1-\sum_{j=1}^{N} p_{i1;j1}^{(r)})/n_i = \pi(\mathbf{n}_{ir}^-)\lambda_{i1}^{(r)}, \tag{2.21}$$

where $\mathbf{n}_{ir}^- \in S^{(R)}$. Equation 2.20 corresponds to the case where a customer of type r moves from node i to node j, or vice versa, while eq. 2.21 corresponds to the case where a customer of type r leaves or enters the network. In the case that routing chain r is closed, we have

$$\sum_{j=1}^{N} p_{i1;j1}^{(r)} = 1$$

and $\lambda_{i1}^{(r)} = 0$, so that only eq. 2.20 need be considered. By substitution of the explicit expression for $\pi(.)$ into eqs. 2.20 and 2.21, it may be verified that the detailed balance conditions hold if and only if

$$\alpha_{i1}^{(r)}p_{i1;j1}^{(r)} = \alpha_{j1}^{(r)}p_{j1;i1}^{(r)} \tag{2.22}$$

and

$$\alpha_{i1}^{(r)}(1-\sum_{j=1}^{N} p_{i1;j1}^{(r)}) = \lambda_{i1}^{(r)}. \tag{2.23}$$

Equation 2.22 implies that the average flow of type r customers from node i to node j is equal to that from node j to node i. Equation 2.23

implies that the average flow of type r customers from node i out of the network is equal to the average flow of exogeneous type r arrivals into node i.

In a similar manner, but at the expense of some additional notation, we may establish the conditions under which reversibility holds in a multiple-chain BCMP queueing network *with* class-switching. In this situation, the conditions are

$$\alpha_{ic}^{(r)} p_{ic;jd}^{(r)} = \alpha_{jd}^{(r)} p_{jd;ic}^{(r)} \tag{2.24}$$

and

$$\lambda_{ic}^{(r)} = \alpha_{ic}^{(r)} \Big(1 - \sum_{j=1}^{N} \sum_{d \in C_{jr}} p_{ic;jd}^{(r)}\Big), \tag{2.25}$$

where c is the class index. It may also be verified that if eqs. 2.24 and 2.25 are satisfied, then reversibility holds when there are LCFSPR or IS queues in the network. If $R = 1$ and there is no class-switching, then reversibility holds when there are FCFS queues if eqs. 2.22 and 2.23 are satisfied. We cannot, however, ensure reversibility if there are FCFS queues and multiple types of customers or class-switching since, in this situation, the state description of the nodes includes the ordering of the customers in the queues and we cannot, in general, ensure that all pairs of states will be doubly-connected or not connected at all.

Although the conditions for reversibility impose severe restrictions on the structure of the routing matrices $\mathbf{P}^{(r)}$ and the magnitudes of the external arrival rates $\lambda_{ic}(r)$, they do allow for a number of network configurations that are of practical importance. These include central-server type networks, two-queue cyclic networks and, more generally, networks whose nodes are arranged in the form of a tree.

In queueing networks that satisfy the above described conditions for reversibility, it is possible to accommodate a certain simple form of *blocking* and maintain a product-form distribution. Since a reversible Markov chain may be truncated to an arbitrary state-space A, $S \supset A$, without altering the form of the equilibrium distribution, we may, for example, exclude all states in which the queue-lengths n_i are above some threshhold, say κ_i [PIT1]. In such a situation, if a customer moves from node j to node i and finds that $n_i = \kappa_i$, then the customer is effectively moved back to node j for another independent round of service. Such a mechanism is known as 'block-and-recirculate' [YAO1]. It is to be noted, however, that the practical value of such blocking mechanisms appears to be limited since they do not give rise to blocking of a genuine kind.

2.3 Applications of Queueing Network Models

Queueing networks of the product-form type have now been used for some time as models in the analysis of computer systems and computer-communication networks. In computer systems, a job or task is represented by a customer and the various resources, including CPUs, memory and I/O devices, are represented by service centers. In models of computer networks, a customer is used to represent a packet or message and the service centers are used to represent the data transmission links together with their buffers. More recently, product-form queueing network models have arisen in a number of other applications including local area networks of workstations and distributed systems of computers.

Product-form queueing networks have also been used as models in a number of more specialized applications including, for example, the analysis of capacity allocation schemes for satellites and buffer management schemes for packet switches. They have also been used as models of circuit-switched communication networks and of multiprocessor systems. Recently, they have also been proposed as models for the analysis of system availability and as performance models for Open Systems Interconnection (OSI) computer-communication protocols [OSI1] which are layered according to the International Standards Organization (ISO) Reference Model [DAY1,STA1].

Apart from having direct applicability in a number of modeling applications such as those described above, product-form queueing networks have also arisen frequently as submodels in larger models that themselves do not support a product-form equilibrium state-distribution. Examples in which such submodels arise include models of time-shared computer systems having a limited number of memory partitions [BRA1,BRA2,JACO1,JACO2,LAZ2], models of computer systems with service centers that do not support a product-form distribution [NEU1,NEU2] and models of computer networks with nested flow-controlled domains [KRZ3,SOU4]. Other more specialized examples that make use of product-form submodels include models to determine admission delays to virtual circuits in computer networks [REI6] and models for the analysis of pacing controls [SCH1].

In the following, we shall overview several of the above mentioned applications and illustrate the applicability of several of the modeling features described in Section 2.1. Survey papers on the use of queueing networks as performance models include [HEI1,KUR1,LAM5,LAZ1,MUN1,REI2,REI5,WON1,WON2].

2.3.1 Queueing Network Models of Computer Systems

A computer system, of the time-sharing type, may be modeled as a closed queueing network of the BCMP type, as illustrated schematically in Figure 2.6. This is the well-known central-server model. There is one customer associated with each 'logged-on' user at the terminals and each customer is assumed to belong to some particular closed routing chain. The customer resides at the

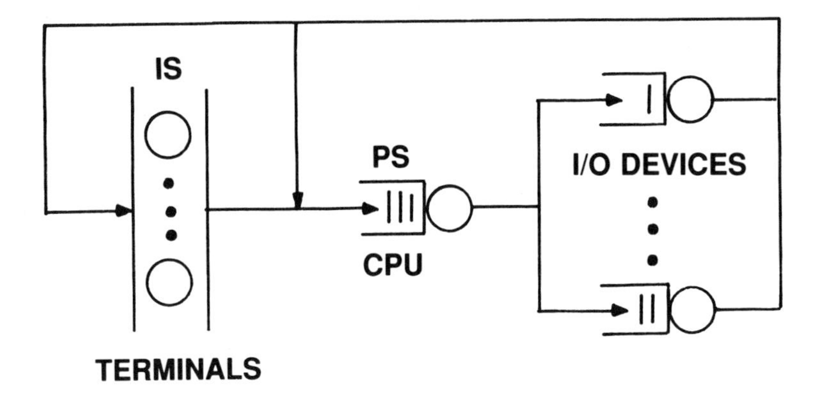

Figure 2.6
Central-Server Model of a Computer System

'terminals' queue during the time that the user is in a 'thinking' or 'typing' state. An IS queue is used to represent the terminals since it is assumed that there is no contention for this resource. When a user submits a job or a task to the system, the customer associated with the user leaves the IS queue and proceeds through a sequence of CPU-I/O device cycles. Once the task is completed, the customer returns back to the IS queue.

By considering the customer associated with each user, or the customers associated with particular groups of users to be of a particular type, we may allow the routing and service-time requirements of a customer to depend on the type of the associated user. In this way, we may model heterogeneous user groups. Using the class-switching feature of BCMP queueing networks, we may specify complex sequences of CPU-I/O accesses to model the behaviour of different types of programs that may, for example, be CPU or I/O bound. We may also allow the routing of a customer among the CPU and I/O device queues to be some finite deterministic sequence. We may also allow for different types of queueing disciplines at the CPU and I/O devices. It is customary to make use of the PS service discipline to approximate round-robin service at the CPU. By making use of multiple open routing chains, in conjunction with the closed routing chains, we may in addition allow for the presence of different types of 'batch' jobs in the system. By making use of the state-dependent arrival mechanism, we may allow the rate of arrival of batch jobs to depend on the number already in the system.

2.3.2 Queueing Network Model of a Distributed Computer System

A model of a distributed system of computing sites may be formulated [GOL1,SOU3] by interconnecting a group of central-server models of the type illustrated in Fig. 2.6. The resulting model is illustrated schematically in Figure 2.7. In this model, we have a number of customers associated with the terminals at each site and we associate a closed routing chain with each such group of users. The customers associated with a particular site place demands on the resources which are local to their site as well as on those at remote sites. In addition, there is a queue for the communication network that links the various sites. In a simplistic model, we would represent the communication network by a FCFS or PS queue. To be more realistic, we can represent a multiple-access communication network of the Ethernet or polling type using a queue with state-dependent service-rates, the parameters of which may be determined *a priori* from an analysis of a detailed submodel of the media-access scheme [ALM1,BUX1,KIN1,LAM6].

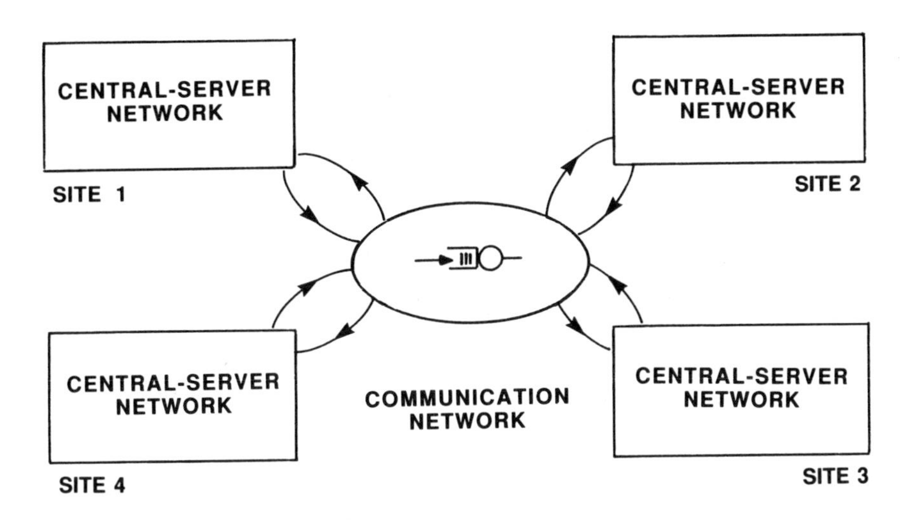

Figure 2.7
Queueing Network Model of a Distributed Computer System

2.3.3 Queueing Network Models of Communication Networks

For the modeling of communication networks, various types of queueing networks can be used, depending on the type of service that is provided in the network. For message-switched networks, and packet-switched networks that provide datagram service, it is customary to adopt an open queueing network model of the Jackson type. An important assumption which is inherent in the adoption of such models for communication networks is that the service-time requirements of a customer at the various service centers along a route are independent random variables. This is known as Kleinrock's *independence assumption* [KLE1]. Clearly, such an assumption is unrealistic since it ignores the usually strong correlation that exists between the service-times of a customer at a

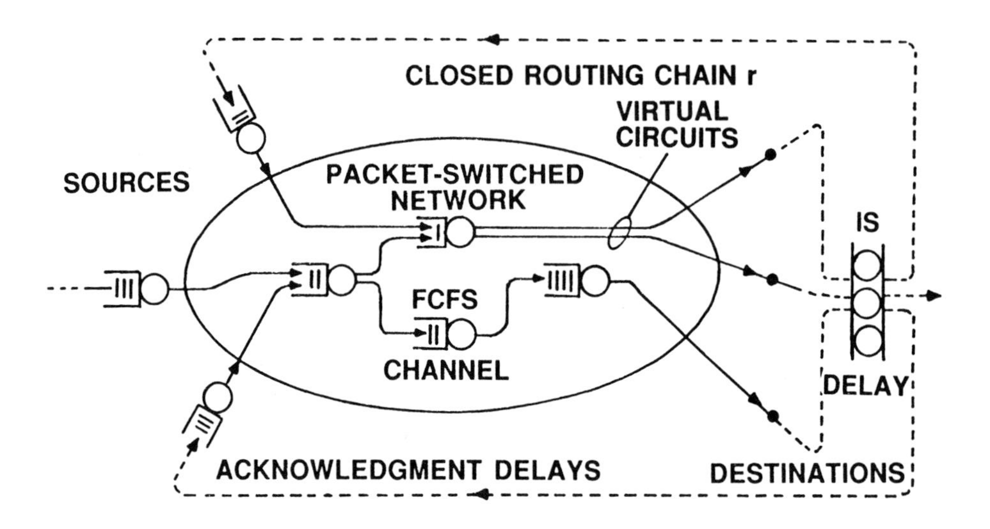

Figure 2.8
Queueing Network Model of a Packet-Switched Network

sequence of service centers. It has been argued, however, that, in practice, the independence assumption usually does not lead to excessive errors. To be more pragmatic, such an assumption must be made in the interest of analytical tractability.

Closed queueing networks of the BCMP type with multiple types of customers are used to model computer-communication networks that provide window flow-controlled virtual-circuit service connections at the transport level [PEN1,REI1,GEO1]. In these models, we associate a closed routing chain r with each virtual-circuit that is assumed to be established across the network. The number of customers of type r, associated with chain r, is set equal to the window size of the connection. Between each source-destination pair, we can use a pair of closed routing chains to model a full-duplex connection. Although acknowledgements are usually piggybacked onto packets moving in the reverse direction, it is customary to simply model the effect of acknowledgement delays by using an IS queue, as illustrated in Figure 2.8. Once a customer has reached its destination, the associated packet is passed to the receiving entity at the destination and the customer proceeds back to the source through the IS queue. The delay through the IS queue effectively models the acknowledgement delay. Once the customer has returned to the source, a new packet is permitted to be injected into the network. An inherent limitation of the model is that, in order for a product-form distribution to hold, it is necessary that it be assumed that the service-time requirements be the same for all types of customers at the FCFS queues used to model the transmission channels of the packet network.

2.3.4 Queueing Model of a Network of Workstations

A queueing network model of a local area network of workstations is illustrated in Figure 2.9. In this model, we associate a queue with each workstation and shared resource (e.g bus, file server, printer) in the network and a closed routing chain with the user at each workstation [BAL2,CON7,CON8]. The number of customers in routing chain r is set equal to the number of tasks that user r may issue concurrently. Each customer requires processing at the workstation to which it is associated, as well as at any of the remote workstations and shared resources in the network. The resulting model is a closed multiple-chain BCMP queueing network similar to that of a distributed system of computers.

2.3.5 Queueing Model of a Circuit-Switched Network

Queueing networks of the product-form type also arise as models of circuit-switched communication networks [BUR1,CON10,DZI1]. Consider a graph with N nodes and E links. The number of calls which

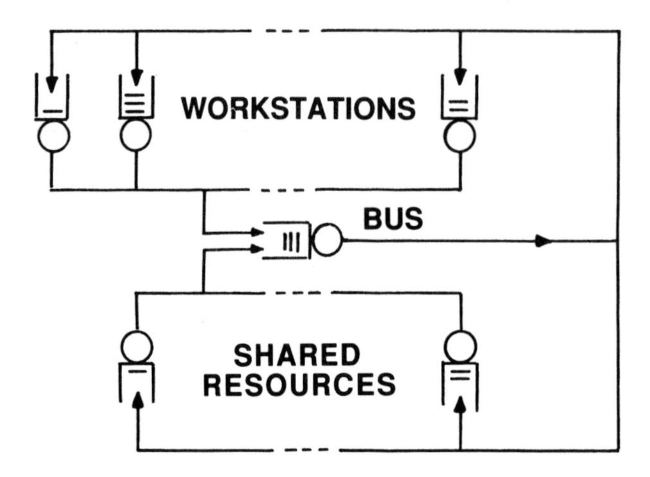

Figure 2.9
Queueing Network Model of a Local Area Network of Workstations

may be accommodated on link e is limited to C_e. There are R types of traffic that may be distinguished by their sources and destinations. Type r traffic consists of calls which are made from node a_r to node b_r. Type r calls arrive according to a Poisson process at rate λ_r and the call holding times are independent and arbitrarily distributed with mean μ_r^{-1}. The state of the system is $\mathbf{n}^{(R)} = (n_1,\ldots,n_R)$, where n_r is the number of calls of type r in progress. The state-space is

$$S_1 = \{\mathbf{n}^{(R)} \mid n_r \geq 0,\ r = 1,\ldots,R;\ \sum_{r=1}^{R} n_r\delta_{re} \leq C_e,\ e = 1,\ldots,E\},$$

where

$$\delta_{re} = \begin{cases} 1, & \text{if type r traffic is routed over link e,} \\ 0, & \text{otherwise.} \end{cases}$$

The above described system may be viewed as a queueing network with a single queue of the IS type, R types of customers and loss functions $\mathbf{L}_r$, of the type described in Subsection 2.1.5, that restrict the state-space to S_1.

2.3.6 Queueing Models of Satellite Channels and Packet Switches

A model closely related to that of the circuit-switched network is one for a time-division multiple-access (TDMA) satellite transmission channel [AEI1,KAU1]. In such a model, we assume that there are R types of users (customers) that arrive according to Poisson processes and that customers of type r require b_r units of bandwidth for a mean transmission time of μ_r^{-1}. The total available channel bandwidth is C. The state of the channel is described by $\mathbf{n}^{(R)} = (n_1,\dots,n_R)$, where n_r is the number of users of type r which are transmitting. The state-space is

$$S_2 = \{\mathbf{n}^{(R)} \mid n_r \geq 0,\ r = 1,\dots,R;\ \sum_{r=1}^{R} n_r b_r \leq C\}.$$

This system may be viewed as a queueing network with a single queue of the IS type, R types of customers and loss functions $\mathbf{L}_r$, of the type described in Subsection 2.1.5, that restrict the state-space to S_2. By altering the loss functions, we may change the policy whereby bandwith is allocated to incoming users.

Another closely related model is that of a packet switch, as illustrated in Figure 2.10 [IRL1,KAM1]. In this model, there are R types of arriving packets (customers) that may be distinguished by their destinations. There is a single dedicated transmission channel towards each of these destinations. Type r packets arrive according to a Poisson process at rate λ_r and their mean service-time

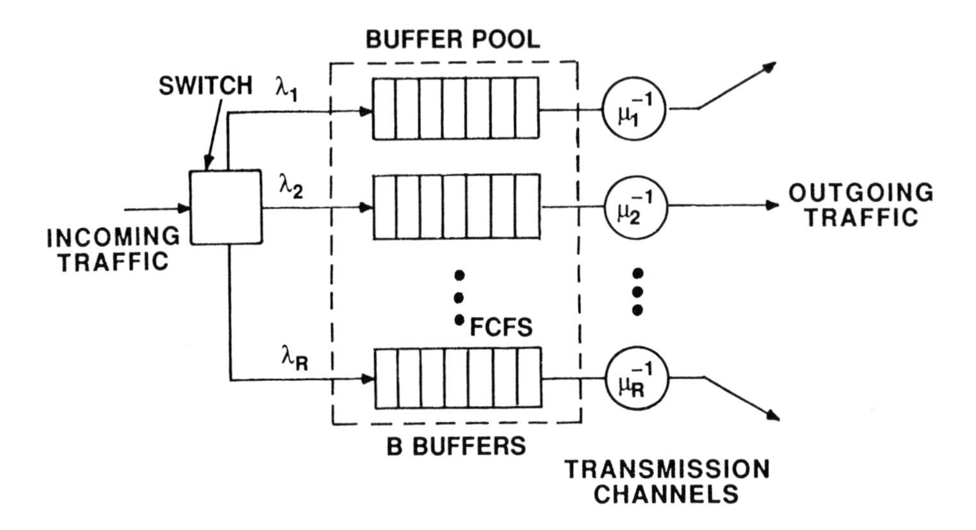

Figure 2.10
Queueing Model of a Packet Switch

requirement (transmission time) is μ_r^{-1}. The total number of available buffers is B. The state is $\mathbf{n}^{(R)} = (n_1,...,n_R)$, where n_r is the number of type r packets in transmission or in the buffers. The state-space is

$$S_3 = \{\mathbf{n}^{(R)} \mid 0 \le n_r \le (B + 1),\ r = 1,...,R;\ \sum_{r=1}^{R} n_r \le (B + R)\}.$$

In this situation, we have a set of R FCFS queues and loss functions $\mathbf{L}_r$ that restrict the state-space to S_3. By altering the form of the loss functions, one may model a variety of different schemes whereby the available buffers are allocated to incoming packets.

2.3.7 Queueing Model of a Multiprocessor System

A particular application in which the MSCCC queueing discipline arises is in the performance modeling of multiprocessor systems [LEB1,MARS2]. Consider a situation in which there are P processors and M memory modules which may be accessed through any one of B buses. When processor r, r = 1,...,P, makes a memory access, we assume that it requests module m with probability q_{mr}. If the particular module is already being accessed by another processor or if there are no buses currently available, then the processor joins a first-in first-out (FIFO) queue. When a bus becomes available, service is given to the first waiting processor in FIFO order that wishes to access a memory module which is not already being accessed by another processor.

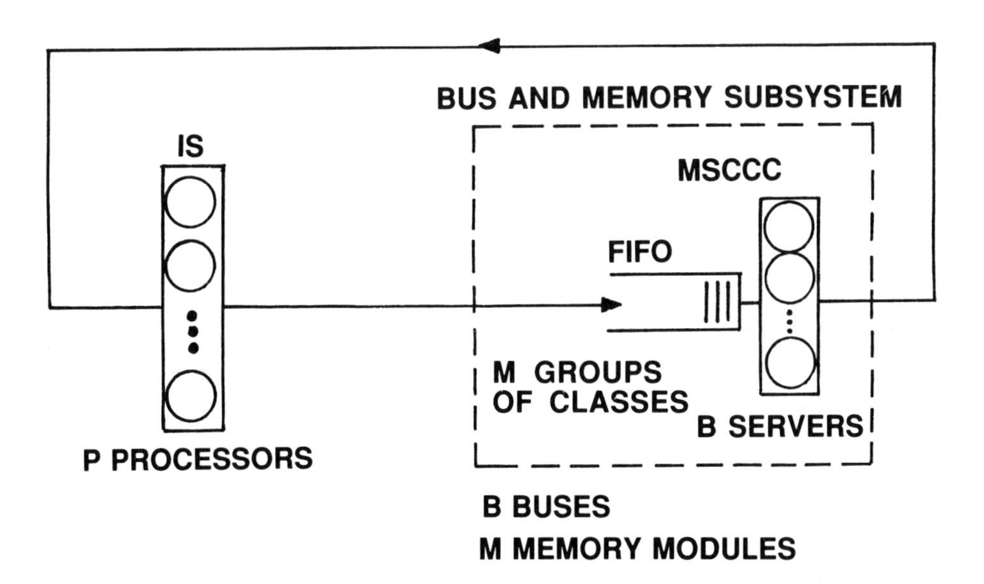

Figure 2.11
Queueing Model of a Multiprocessor System

The system described above may be represented by a product-form queueing network with one IS queue and one queue of the MSCCC type having B servers and M groups of classes, as illustrated in Figure 2.11. In this model, we have P customers, corresponding to the P processors, which are all considered to be of a different type. With each customer we may associate a separate closed routing chain r. During the time that a processor is executing instructions, the associated customer resides in the IS queue. When a memory access is required at module m, the customer moves into the MSCCC queue and joins group m with probability q_{mr}.

2.3.8 System Availability Models

In certain circumstances, product-form queueing networks may also be used as computer system availability models [GOY1,GOY2]. In such models, a system is considered to be made up of a number of hardware and software components that are either in an operational or a failed condition. When a component fails, it joins a repair queue, or one of a number of repair queues, for service. The overall state of the system depends on the conditions of the components. If certain combinations of the components are 'up', then the system itself is considered to be operational. The availability of the system is defined as the proportion of time, in equilibrium, that the *system* is found to be operational.

A queueing network model for system availability is illustrated in Figure 2.12. It is essentially a generalization of the classical machine-repair model [LAV2]. In this model, we associate a customer with each component in the system. With each such customer, we can associate a distinct closed routing chain r. A customer is either at the failure queue, where it resides during the time that the associated component is operational, or it is at one of the repair queues. An IS queue is usually used to represent the failure process. This implies that components fail independently of one another. At the repair queues, we may have different service disciplines such as, for example, FCFS or random order service (ROS) (ROS is equivalent to PS, as far as the marginal state-distribution is concerned [KEL3]). Assuming that there are no failure or repair dependencies [GOY2] among the various components that would preclude a product-form state-distribution, the distribution for the system assumes the form of eq. 2.10. The system availability A is given by

$$A = \sum_{\mathbf{n}^{(R)} \in S_4} I(\mathbf{n}^{(R)})\pi(\mathbf{n}^{(R)}),$$

where S_4 is the state-space of the model, $\mathbf{n}^{(R)}$ is the state of the failure/repair model and

$$I(\mathbf{n}^{(R)}) = \begin{cases} 1, & \text{if } \mathbf{n}^{(R)} \text{ is an operational state,} \\ 0, & \text{otherwise.} \end{cases}$$

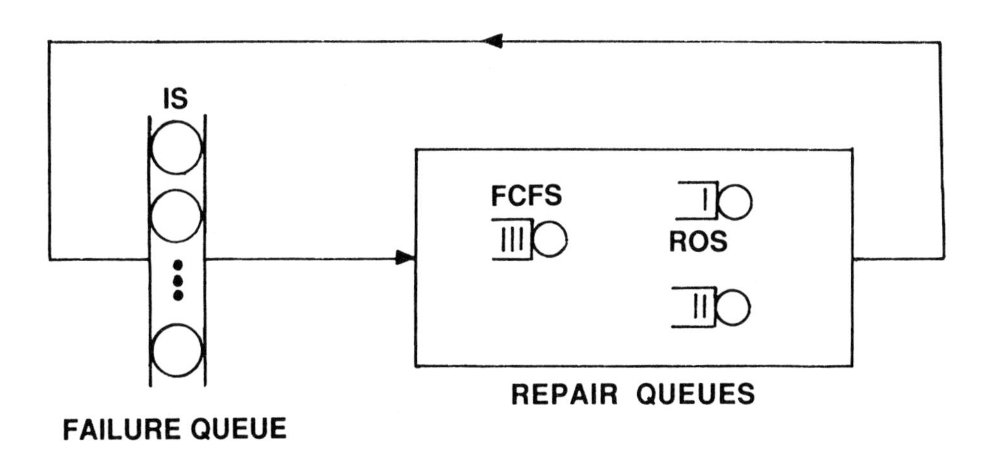

Figure 2.12
Queueing Network Model for System Availability

2.3.9 Queueing Model for OSI Communication Architectures

Recently [KRI1], multiple-chain closed queueing networks of the BCMP type have also been proposed as models for the performance evaluation of OSI communication architectures [OSI1] which are layered according to the ISO Reference Model [DAY1,STA1]. In the seven-layer Reference Model, each layer makes use of the communication services provided by the underlying layer to provide an enhanced set of services to the layer lying above. Associated with each layer are so-called *entities* that correspond to finite-state protocol machines. Entities at adjacent layers communicate via conceptual service-access points through which protocol data units (PDUs) pass. At the inputs to each entity, one may have queues to

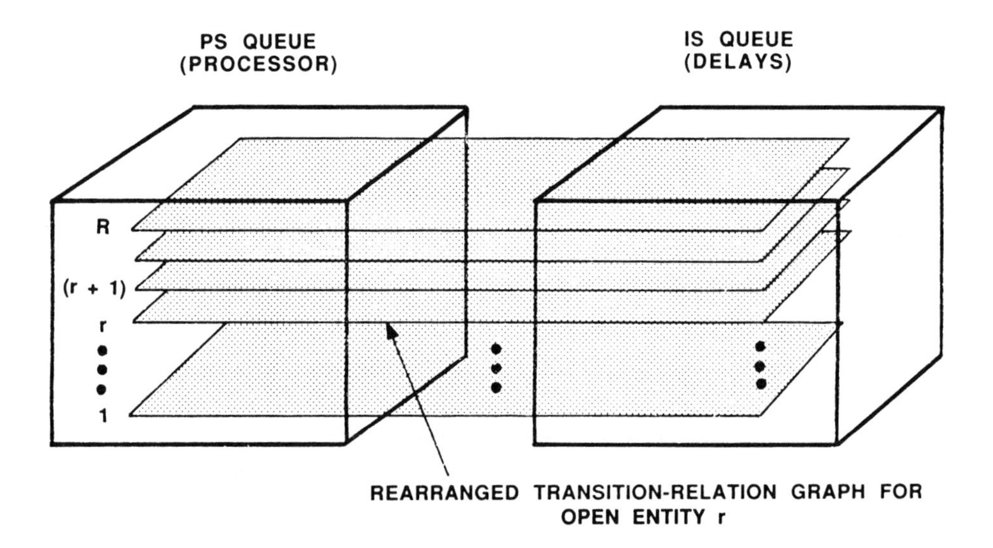

Figure 2.13
Queueing Model of a Layered OSI Communication Architecture

buffer incoming PDUs. Associated with each entity is a peer entity, or possibly a number of peer entities, that may be geographically separated and which intercommunicate via the services of the underlying layer. The peer entities cooperate according to a defined protocol to achieve the functions that are required to realize the service that is to be provided to the layer lying above. At any point in time, an entity has a certain specific state and, depending on the objects (PDUs) that are found in the input queues to the entity, the entity changes to another state according to a defined protocol state-and-action table and certain defined objects are passed to the neighbouring entities residing in the adjacent layers [OSI1].

Kritzinger considers the transition-relation graph (which is derived from the state-and-action table) of the entities in the various layers and distinguishes between those states of the graph that involve only local processing and those which depend on inputs which are received from the environment of the entity such as, for example, Confirm and Indication primitives. The former type of state is termed *active*, while the latter is termed *passive*.

In the model of Kritzinger, a closed routing chain with a single customer is associated with the transition-relation graph of an entity in a particular layer. A distinct class is associated with each possible state so that the movement of the customer through the various classes of the routing chain corresponds to the trajectory of the entity through the possible states of the transition-relation graph. An IS queue is used to model the random waiting delays associated with the passive states and a processing queue of the PS type is used to model the processing delays associated with the active states. When a customer changes class, it moves to the IS queue if the state associated with the next class is passive or it moves to the PS queue if it is of the active type.

In the full model of an open system made up of a number of layers of protocols, there is a closed routing chain, of the type described above, associated with each entity in every layer of the

ISO Reference Model that is in an open state (in the data transfer phase of a connection). The overall multiple-chain closed queueing network model is illustrated schematically in Figure 2.13. It is to be noted that, in view of the results for BCMP queueing networks summarized in Subsection 2.1.3, the analysis of this multiple-chain queueing network model with class-switching is reducible to that of a queueing network with no class-switching.

CHAPTER 3

Decomposition Methods in Queueing Network Analysis

For the analysis of large scale and complex systems with large state-spaces, it is quite natural to consider the application of decomposition methods in an attempt to reduce the inherent analytical and computational difficulties. The fundamental formal theoretical foundation of decomposition methods for the analysis of queueing networks in equilibrium may be attributed to the work of Simon and Ando [SIM1] in the early 1960's. They proposed a so-called *Decomposition and Aggregation* technique for the analysis of systems that exhibit certain properties of *near-complete decomposability.* In such systems, it is assumed that there exist a number of interacting subsystems such that the interactions taking place within each particular subsystem are strong relative to those taking place between the individual subsystems themselves. The theory developed by Simon and Ando was motivated by the modeling of economic systems. The method, as it applies to Markov chains in general, assumes that there exist groups of communicating states such that the coupling among the states belonging to the same group is strong while that between states of different groups is weak. The primary feature of the Decomposition and Aggregation technique is to reduce the analysis of a large system into that of a set of smaller problems [COU5].

Although the Decomposition and Aggregation technique, in general, yields approximate results, for systems which are nearly-completely decomposable the error incurred is of the order of the so-called *maximum degree of coupling* ε, in the sense that the error tends to zero in ε [COU3]. The quantity ε is a measure of the amount of coupling that actually exists between the subsystems. For systems which do not satisfy the technical conditions of near-complete decomposability, the Decomposition and Aggregation technique, in general, yields approximate results involving errors whose magnitudes are not easily quantified. Nevertheless, in such circumstances, the method is still of considerable practical use. A large problem may be reduced into a set of smaller ones and the accuracy of the approximation may still be evaluated by comparing the results obtained with empirical evidence in the form of exact results obtained by 'brute force' or estimates obtained by Monte Carlo simulation.

In certain circumstances, particularly for product-form queueing networks, the Decomposition and Aggregation procedure can provide exact results. In such cases, the value of the Decomposition and Aggregation methodology lies not so much in the numerical results that the method may be used to provide but, rather, in the theoretical insight which is provided concerning the structure of the problem at hand. As is to be seen in this and the following chapter, this insight proves to be of considerable value in understanding the structure of queueing network problems. Indeed, in the developments to be made here, our main use of the Decomposition and Aggregation methodology of Simon and Ando will be as a theoretical formalism rather than as a computational technique.

In the following, we shall first consider the use of the general Decomposition and Aggregation procedure of Simon and Ando, as it applies to Markov chains in general, as a means to obtain the equilibrium distribution of a large system efficiently. The general conditions are reviewed under which the method is guaranteed to provide exact results. In Sections 3.2 to 3.5, we consider the application of the method to product-form queueing networks. It is shown how the method leads to multiple-level hierarchies of interrelated subsystems that may be identified in a queueing network. We also show how the method provides a basis for the parametric analysis of queueing networks. In Section 3.6, it is proved that the general Decomposition and Aggregation technique invariably yields exact results for reversible queueing networks. This result provides the basis for parametric analysis techniques for product-form queueing networks that may be applied to *arbitrary* subsystems of interest. Such generalized parametric analysis techniques are considered in Section 3.7. The developments to be made in this chapter will lead us quite naturally towards the general theory for the construction of computational algorithms to be formulated in Chapter 4.

3.1 Simon and Ando Decomposition and Aggregation

Consider an irreducible infinitesimal generator matrix $\mathbf{Q}$, of order n, associated with an homogeneous, continuous-time Markov chain with state-space S and suppose that $\mathbf{Q}$ can be partitioned into submatrices, as illustrated in Figure 3.1, such that the set of states $S(\alpha)$, $S \supset S(\alpha)$,

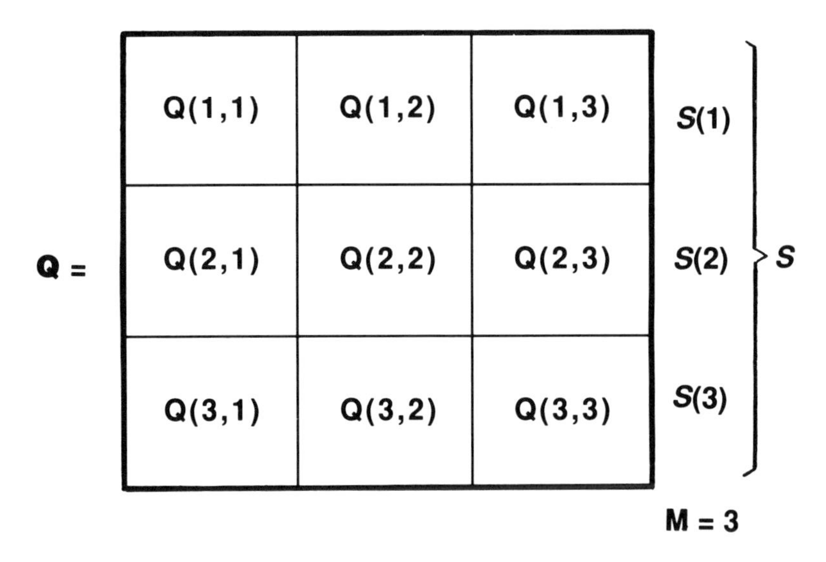

Figure 3.1
Partitioning of the Infinitesimal Generator Matrix $\mathbf{Q}$

associated with the diagonal block α, $1 \le \alpha \le M$, is a group of communicating states. Let $n(\alpha)$ be the order of the square diagonal block $\mathbf{Q}(\alpha,\alpha)$. We use the notation $q_{ij}(\alpha,\beta)$ to denote the *(i,j)*th element of the submatrix $\mathbf{Q}(\alpha,\beta)$. Also let $s_{i\alpha}$ denote the off-diagonal block row-sum associated with the *i*th row of $\mathbf{Q}(\alpha,\alpha)$. Formally,

$$s_{i\alpha} = \sum_{\substack{\beta=1 \\ \beta \neq \alpha}}^{M} \sum_{j=1}^{n(\beta)} q_{ij}(\alpha,\beta).$$

Furthermore, suppose without loss of generality, that $\mathbf{Q}$ is normalized so that

$$\text{Max}\{ \sum_{\substack{j=1 \\ j \neq i}}^{n} q_{ij} \mid i = 1,\ldots,n\} = 1,$$

where q_{ij} is the *(i,j)*th element of $\mathbf{Q}$. With this normalization, the matrix $(\mathbf{Q}+\mathbf{I})$, where $\mathbf{I}$ is an identity matrix of order n, is a stochastic matrix.

A left eigenvector, associated with the eigenvalue one of the matrix $(\mathbf{Q}+\mathbf{I})$, is denoted by π and partitioned as $\pi = (\pi_1,\ldots,\pi_M)$, where $\pi_\alpha = (\pi_{1\alpha},\ldots,\pi_{n(\alpha)\alpha})$, so as to conform with the partitioning of $\mathbf{Q}$, as has been adopted in Fig. 3.1. In a similar manner, let $s = (s_1,\ldots,s_M)$, where $s_\alpha = (s_{1\alpha},\ldots,s_{n(\alpha)\alpha})$.

Assuming that the strength of the coupling between the states of each particular subset $S(\alpha)$ is strong relative to that between states of different subsets, it is quite natural to consider writing $\mathbf{Q}$ in the form

$$\mathbf{Q} = \mathbf{Q}^* + \varepsilon\, \mathbf{C},$$

where $\mathbf{Q}^* = \mathbf{diag}(\mathbf{Q}_1^*,...,\mathbf{Q}_M^*)$, $\mathbf{Q}_\alpha^*$ is an infinitesimal generator matrix constructed from $\mathbf{Q}(\alpha,\alpha)$ by distributing the quantity $s_{i\alpha}$ over the non-zero elements of the *i*th row of $\mathbf{Q}(\alpha,\alpha)$, for $1 \leq i \leq n(\alpha)$, and

$$\varepsilon = \text{Max } \{s_{i\alpha} \mid 1 \leq \alpha \leq M, 1 \leq i \leq n(\alpha)\}.$$

For the types of systems under consideration, the quantity ε, usually referred to as the *maximum degree of coupling* between the subsystems $\mathbf{Q}_\alpha^*$, $1 \leq \alpha \leq M$, is relatively small compared to the off-diagonal elements of the matrices $\mathbf{Q}_\alpha^*$, $1 \leq \alpha \leq M$. If ε is sufficiently small, then $\mathbf{Q}$ is said to be *nearly-completely decomposable.*

In a set of theorems proved by Simon and Ando [SIM1], it has been established that if ε is sufficiently small, then the dynamic behaviour of the transient state-distribution $\pi(t)$, where $\pi(t) = (\pi_1(t),...,\pi_M(t))$, as a function of time t, may be seen to evolve through four distinct phases. In the first phase, called *short-term dynamics*, $\pi_\alpha(t)$ and $\pi_\alpha^*(t)$ evolve similarly, where $\pi_\alpha^*(t)$ is the transient state-distribution associated with $\mathbf{Q}_\alpha^*$. It is during this particular phase that the transient effects due to the strong couplings within the subsystems $\mathbf{Q}(\alpha,\alpha)$, $1 \leq \alpha \leq M$, are felt. In the second phase, called *short-term equilibrium*, $\pi_\alpha(t)$ and $\pi_\alpha^*(t)$ reach a similar equilibrium in the sense that the forms of the distributions $\pi_\alpha(t)$ and $\pi_\alpha^*(t)$ become similar to one another. In the third phase, known as *long-term dynamics*, the transient effect due to the weak coupling between the subsystems $\mathbf{Q}(\alpha,\alpha)$, $1 \leq \alpha \leq M$, is felt. During this phase, $\pi_\alpha(t)$ and $\pi_\alpha^*(t)$ continue to move towards their equilibrium values, respectively, and the relative values among the state probabilities $(\pi_{1\alpha}(t),...,\pi_{n(\alpha)\alpha}(t))$ and $(\pi_{1\alpha}^*(t),...,\pi_{n(\alpha)\alpha}^*(t))$ are maintained approximately. The marginal distribution $\Pi(t)$, where $\Pi(t) = (\Pi_1(t),...,\Pi_M(t))$ and

$$\Pi_\alpha(t) = \sum_{i=1}^{n(\alpha)} \pi_{i\alpha}(t),$$

evolves through a long-term transient. Finally, in the last phase, known as *long-term equilibrium*, both $\Pi(t)$ and $\pi(t)$ move towards their respective equilibrium values and the overall equilibrium condition is reached.

In order for a Markov chain to exhibit the 'nearly-completely decomposable' behaviour described above, it is *necessary* that ε be sufficiently small. A necessary condition, as given in [COU1, Section 3.1], is that

$$\min_{\alpha} | \lambda(1_\alpha) | > \max_{\alpha} | \lambda^*(2_\alpha) |, \tag{3.1}$$

where $\alpha = 1,\ldots,M$, $\lambda(1_\alpha)$ is the largest eigenvalue of $(\mathbf{Q}(\alpha,\alpha) + \mathbf{I})$ and $\lambda^*(2_\alpha)$ is the second largest eigenvalue of $(\mathbf{Q}_\alpha{}^* + \mathbf{I})$. However, in eq. 3.1, none of the terms is known *a priori*. In view of this difficulty, assuming that $\varepsilon < 1/2$, Courtois [COU1, Section 3.2] has formulated the following *sufficient* condition for near-complete decomposability:

$$\varepsilon < (1 - \max_{\alpha} | \lambda^*(2_\alpha)|) / 2. \tag{3.2}$$

The more relaxed sufficient condition

$$\varepsilon < \max_{\alpha} | \lambda^*(2_\alpha)| / 2 \tag{3.3}$$

has subsequently been reported in [BALS1]. Equations 3.2 and 3.3 may be useful in practice since expressions for upper bounds on $|\lambda^*(2_\alpha)|$ are known (see [COU1, Section 2.2.2]) and for certain matrices, such as those which are tri-diagonal, closed-form expressions for the eigenvalues may be obtained analytically (see [COU1, Appendix 4]).

The four phases that may be identified in the dynamic behaviour of a nearly-completely decomposable Markov chain suggests the following state-space decomposition technique for obtaining an approximation $\mathbf{z}$ for the equilibrium distribution π. The technique consists of two main steps, namely, 'decomposition' and 'aggregation', and proceeds as follows:

Simon and Ando Decomposition and Aggregation:

STEP 1: (*Decomposition*) Construct $\mathbf{Q}^*$ from $\mathbf{Q}$ and find $\mathbf{v}^*$, where $\mathbf{v}^*(\mathbf{Q}^*+\mathbf{I}) = \mathbf{v}^*$, $\mathbf{v}^* = (\mathbf{v}_1^*,...,\mathbf{v}_M^*)$ and $\mathbf{v}_\alpha^* = (v_{1\alpha}^*,...,v_{n(\alpha)\alpha}^*)$, subject to the constraint that $\mathbf{v}_\alpha^*\mathbf{1}^T = 1$, for $1 \leq \alpha \leq M$, where $\mathbf{1}^T$ is a compatible column vector, all of whose elements are one.

STEP 2: (*Aggregation*) Construct the infinitesimal generator $\mathbf{A}$, of order M, of the so-called *reduced system*, as follows. With a_{ij} as the *(i,j)*th element of $\mathbf{A}$, let

$$a_{ij} = \begin{cases} \mathbf{v}_i^*\mathbf{Q}(i,j)\mathbf{1}^T, \text{ for } i \neq j, \\ -\sum_{\substack{k=1 \\ k \neq i}}^{M} a_{ik}, \text{ for } i = j. \end{cases} \tag{3.4}$$

Find the equilibrium distribution $\mathbf{u}$ of the reduced system $\mathbf{A}$, where $\mathbf{u}(\mathbf{A}+\mathbf{I}) = \mathbf{u}$ and $\mathbf{u}\mathbf{1}^T = 1$.

STEP 3: Compute the approximation for π as $\mathbf{z} = [[u_1\mathbf{v}_1^*],...,[u_M\mathbf{v}_M^*]]$.

In the above, the Decomposition step may be interpreted as the one where we find an approximation $\mathbf{v}_\alpha^*$ to the true conditional distribution $G_\alpha^{-1}\pi_\alpha$, where $G_\alpha = \sum_{i=1}^{n(\alpha)} \pi_{i\alpha}$ is simply a normalization constant. The distribution $G_\alpha^{-1}\pi_\alpha$ is, of course, unknown *a priori*. The Aggregation step may be interpreted as the one where we find an approximation $\mathbf{u}$ for the true marginal equilibrium distribution Π, where $\Pi = (\Pi_1,...,\Pi_M)$ and $\Pi_\alpha = \sum_{i=1}^{n(\alpha)} \pi_{i\alpha}$, which is also unknown *a priori*.

The second step is known as an Aggregation step since, in effect, we are finding the equilibrium distribution for a 'reduced' system in which all of the states belonging to each particular group $S(\alpha)$, $1 \le \alpha \le M$, in the original system are aggregated, or reduced, into one single state. The quantity a_{ij}, $i \ne j$, may be interpreted as the exact average transition rate, or *probability flux*, from $S(i)$ to $S(j)$, if $\mathbf{v}_i^*$ were actually equal to π_i. Finally, in Step 3, the approximate conditional distributions $\mathbf{v}_\alpha^*$, $1 \le \alpha \le M$, are deconditioned using the approximate marginal distribution $\mathbf{u}$ to obtain the approximation $\mathbf{z}$ for π.

The Decomposition and Aggregation technique defined above may be extended quite naturally into a hierarchical Decomposition and Aggregation technique if any of the subsystems $\mathbf{Q}_\alpha^*$, $1 \le \alpha \le M$, are themselves nearly-completely decomposable. It is also conceivable that the reduced system $\mathbf{A}$ may, itself, be nearly-completely decomposable.

As can be seen from Step 1, the construction of $\mathbf{Q}^*$, by distributing the quantities $s_{i\alpha}$, $1 \le i \le n(\alpha)$, over the rows of $\mathbf{Q}(\alpha,\alpha)$, where $1 \le \alpha \le M$, is unspecified and left open to us. The resulting error vector $(\mathbf{z}-\pi)$, however, depends completely on the manner in which this construction is carried out.

Given that the error incurred depends on the manner in which we construct $\mathbf{Q}^*$, it is useful to understand the theoretical conditions under which it is guaranteed that exact results will be obtained. The following theorem, due to Courtois [COU2] (see also [COU4]), establishes the necessary and sufficient conditions under which $\mathbf{z}$ is exactly equal to π, regardless of the actual value of the degree of coupling ε that exists in the system.

Theorem 3.1: The Decomposition and Aggregation procedure, as defined above, yields the exact equilibrium distribution π for the continuous-time Markov chain with infinitesimal generator $\mathbf{Q}$ if, and only if, $\mathbf{Q}^*$ is constructed in such a way that for $1 \le \alpha \le M$, the distribution $\mathbf{v}_\alpha{}^*$ is actually equal to the true conditional distribution $G_\alpha{}^{-1}\pi_\alpha$.

Although Theorem 3.1 does not provide the means to ensure that exact results are obtained, it does suggest the direction to be followed in an attempt to minimize the error that will be incurred. Given that one may have some insight concerning the structure of the problem at hand, this may sometimes be used in an attempt to construct the subsystems $\mathbf{Q}_\alpha{}^*$ in such a way that the conditional distributions $\mathbf{v}_\alpha{}^*$ may actually be quite close to $G_\alpha{}^{-1}\pi_\alpha$. The subject of constructing $\mathbf{Q}^*$ to minimize the resulting error has been considered in [COU4].

As has been mentioned, the Decomposition and Aggregation procedure is also of practical use for the analysis of systems that do not satisfy necessarily the technical conditions of near-complete decomposability, as expressed by eq. 3.1. Indeed, the method, and closely related variants, have been applied as approximation techniques in a variety of practical applications such as, for example, the modeling of virtual memory and multiprogramming, concurrency and database locking. Although the variety of approximate solution techniques that have been proposed for these and other problems have not always been formulated directly in terms of the Simon and Ando Decomposition and Aggregation theory, the concepts which have been employed are equivalent fundamentally, the pervasive concept being that of analyzing a set of conditional subsystems and a reduced system. A unified view of approximation techniques, as they may be related to the method of Decomposition and Aggregation, has been presented by Brandwajn [BRA3].

Finally, it is to be mentioned that since the development of the original Simon and Ando Decomposition and Aggregation technique and the subsequent work of Courtois in this area, a number of other related solution techniques have been developed. These include iterative decomposition and aggregation techniques [CAO1] and so-called 'bounded-aggregation' methods [COU6] (see also [COU7]).

3.2 Decomposition by Service Center

The application of the Simon and Ando technique to the analysis of queueing networks of the product-form type was originally carried out by Courtois [COU1,COU2,COU8], who adopted what may be called a *Decomposition by Service Center* approach. Courtois originally applied the Decomposition and Aggregation methodology to the class of Gordon-Newell queueing networks [COU8] and subsequently extended the applicability of the technique to closed, multiple-chain BCMP queueing networks in [COU2]. For the sake of simplicity, we shall first consider the technique of Courtois, as it applies to a Gordon-Newell network, assuming that there is a single fixed-rate server at each node in the network ($\mu_i(n_i) = 1$). In the following, we shall denote a closed multiple-chain BCMP queueing network with N nodes and a population vector $\mathbf{K}$ by $B(N,\mathbf{K})$, where $\mathbf{K} = (K_1,...,K_R)$. With this notation, a Gordon-Newell network is denoted by $B(N,K_1)$, where K_1 is the population of the single, closed routing chain contained in the network.

Consider Figure 3.1 and suppose that $\mathbf{Q}$ is the infinitesimal generator corresponding to the queueing network $B(N,K_1)$. Courtois decomposes the state-space S associated with $\mathbf{Q}$ according to the number of customers at some particular node in the network, say the Nth node, without loss of generality. Let the set of states $S(\alpha)$ consists of those states $\mathbf{n}$, $\mathbf{n} = (n_1,...,n_N)$, of the Gordon-Newell network for which $n_N = \alpha$, where n_N is the number of customers at node N. Formally,

$$S(\alpha) = \{\mathbf{n} \mid \mathbf{n} \in S, n_N = \alpha\}.$$

The number of partitions in the decomposition is $M = K_1 + 1$ since $0 \le n_N \le K_1$. This type of decomposition, for the particular

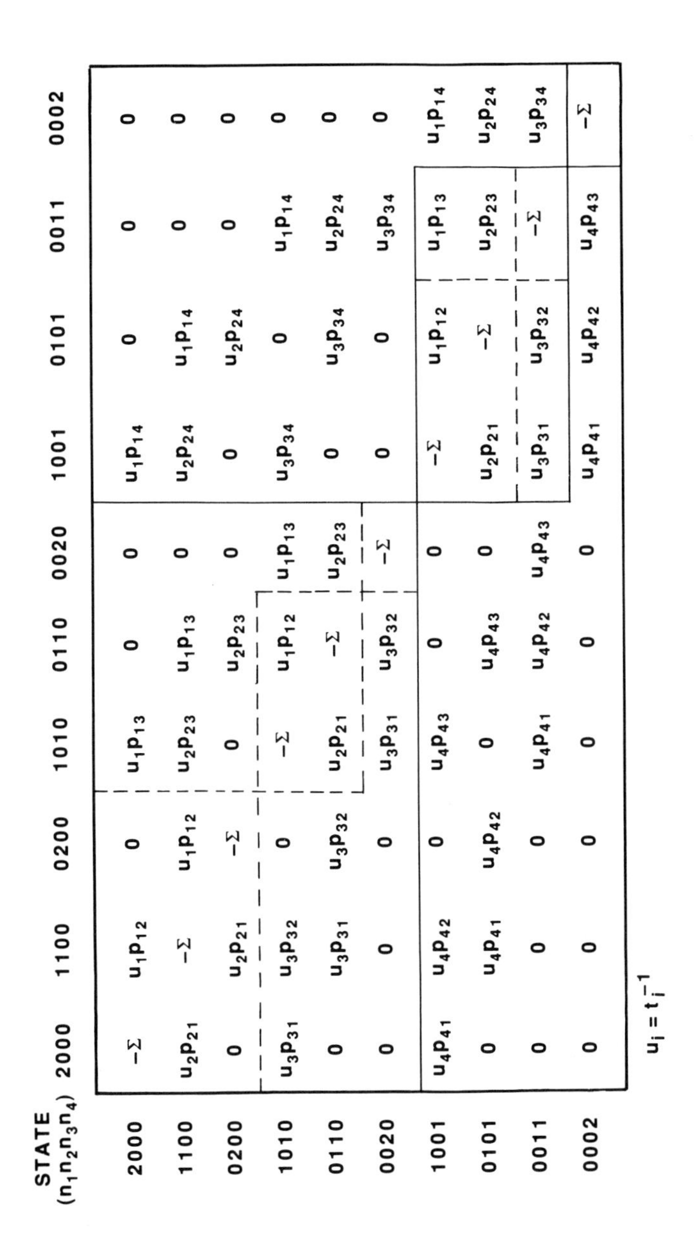

STATE ($n_1n_2n_3n_4$)	2000	1100	0200	1010	0110	0020	1001	0101	0011	0002
2000	$-\Sigma$	u_1p_{12}	0	u_1p_{13}	0	0	u_1p_{14}	0	0	0
1100	u_2p_{21}	$-\Sigma$	u_1p_{12}	u_2p_{23}	u_1p_{13}	0	u_2p_{24}	u_1p_{14}	0	0
0200	0	u_2p_{21}	$-\Sigma$	0	u_2p_{23}	0	0	u_2p_{24}	0	0
1010	u_3p_{31}	u_3p_{32}	0	$-\Sigma$	u_1p_{12}	u_1p_{13}	u_3p_{34}	0	u_1p_{14}	0
0110	0	u_3p_{31}	u_3p_{32}	u_2p_{21}	$-\Sigma$	u_2p_{23}	0	u_3p_{34}	u_2p_{24}	0
0020	0	0	0	u_3p_{31}	u_3p_{32}	$-\Sigma$	0	0	u_3p_{34}	0
1001	u_4p_{41}	u_4p_{42}	0	u_4p_{43}	0	0	$-\Sigma$	u_1p_{12}	u_1p_{13}	u_1p_{14}
0101	0	u_4p_{41}	u_4p_{42}	0	u_4p_{43}	0	u_2p_{21}	$-\Sigma$	u_2p_{23}	u_2p_{24}
0011	0	0	0	u_4p_{41}	u_4p_{42}	u_4p_{43}	u_3p_{31}	u_3p_{32}	$-\Sigma$	u_3p_{34}
0002	0	0	0	0	0	0	u_4p_{41}	u_4p_{42}	u_4p_{43}	$-\Sigma$

$u_i = t_i^{-1}$

Figure 3.2
Decomposition by Service Center of the Queueing Network B(4,2) (from P.J. Courtois, *Decomposability: Queueing and Computer System Applications*, Academic Press, New York, 1977, with the permission of P.J. Courtois and Academic Press)

system B(4,2), is exemplified in Figure 3.2. The bold lines correspond to the particular state-space decomposition that has been adopted. Assuming this method of decomposition, we now consider the details of the Simon and Ando technique, as specified in Section 3.1.

Consider Step 1, the *Decomposition* step. If we now construct $\mathbf{Q}_\alpha^*$ from $\mathbf{Q}(\alpha,\alpha)$ by distributing the off-diagonal block row-sum $s_{i\alpha}$ over the non-zero elements of the *i*th row of $\mathbf{Q}(\alpha,\alpha)$, then it is seen that $\mathbf{Q}_\alpha^*$ is the infinitesimal generator for a system that may be characterized as a single-chain, closed queueing network with a state-space which is the same as that for a Gordon-Newell network $B(N-1,K_1-\alpha)$ with (N-1) nodes and population $(K_1-\alpha)$. We denote this system constructed from $\mathbf{Q}(\alpha,\alpha)$ as $B^*(N-1,K_1-\alpha)$. The cardinality of the state-space of $B^*(N-1,K_1-\alpha)$ is

$$n(\alpha) = \binom{K_1-\alpha+N-2}{N-2}.$$

The service-requirements t_i and the routing probabilities p_{ij}, $1 \le i,j \le N-1$, associated with $B^*(N-1,K_1-\alpha)$ may, however, be different from those associated with $B(N,K_1)$. The actual service-requirements and routing probabilities in $B^*(N-1,K_1-\alpha)$ depend on how we choose to distribute the quantities $s_{i\alpha}$ when we construct $\mathbf{Q}_\alpha^*$. If this is done in an arbitrary way, then only approximate results will, in general, be obtained by the decomposition and aggregation procedure and, furthermore, the system $B^*(N-1,K_1-\alpha)$ may not even exhibit a product-form equilibrium distribution. We shall, however, see a simple way of carrying out this step and maintaining exact results.

Having carried out the single-level Decomposition step, we now turn our attention to the construction and solution of the reduced system **A** that arises in the *Aggregation* step. The state-space associated with **A** is $\{\alpha \mid 0 \le \alpha \le K_1\}$. This is a direct consequence of the method of decomposition that has been adopted. By inspection of Fig. 3.2, or formally utilizing eq. 3.4, it is seen (see for example [VAN1]) that

$$a_{ij} = \begin{cases} t_N^{-1}(1-p_{NN}), & \text{if } j = i-1, \\ \sum_{x=1}^{N-1} t_x^{-1} p_{xN} \sum_{\substack{\mathbf{n}^{(N-1)} \in S(K_1-i) \\ n_x > 0}} v^*(\mathbf{n}^{(N-1)}, K_1-i), & \text{if } j = i+1, \\ -\sum_{\substack{k=0 \\ k \neq i}}^{K_1} a_{ik}, & \text{if } j = i, \\ 0, & \text{otherwise}, \end{cases} \quad (3.5)$$

where t_i is the mean service-requirement for a customer at node i in $B(N,K_1)$, p_{ij} is the routing probability from node i to node j in $B(N,K_1)$, $\mathbf{n}^{(N-1)} = (n_1,...,n_{N-1})$, n_x is the number of customers at node x in $B^*(N-1,K_1-i)$ and $v^*(\mathbf{n}^{(N-1)},K_1-i)$ is the probability of state $\mathbf{n}^{(N-1)}$ in $B^*(N-1,K_1-i)$. By definition, in the case at hand, the quantity

$$\sum_{\substack{\mathbf{n}^{(N-1)} \in S(K_1-i) \\ n_x > 0}} v^*(\mathbf{n}^{(N-1)}, K_1-i)$$

is simply the utilization $U_x^*(K_1-i)$ of the single-server at node x, $1 \leq x \leq N-1$, in $B^*(N-1,K_1-i)$. The reduced system with infinitesimal generator $\mathbf{A}$ is, therefore, simply a birth-death process which may be interpreted as being a FCFS single-server M/M/1 type queue, as illustrated schematically in Figure 3.3, with a service-time of $t_N/(1-p_{NN})$ and a state-dependent arrival rate

$$\lambda(\alpha) = \sum_{x=1}^{N-1} t_x^{-1} p_{xN} U_x^*(K_1-\alpha),$$

where α is the population of the FCFS queue. The state-distribution $\mathbf{u}$, corresponding to $\mathbf{A}$, is $\mathbf{u} = (u_0,...,u_{K_1})$, where u_k is the probability of k customers being at the FCFS queue,

$$u_k = C^{-1} \prod_{\alpha=1}^{k} \lambda(\alpha-1)t_N/(1-p_{NN}) \qquad (3.6)$$

and C is a normalization constant (in eq. 3.6, an empty product is assumed to take on the value one). Alternatively, we may interpret the system **A** as a two-queue cyclic network with total population K_1, as also illustrated in Fig. 3.3, where the first queue is a single-server FCFS queue with constant service-times of $t_N/(1-p_{NN})$, just as in the M/M/1 type queue considered above, and the second queue is a single-server FCFS queue with unit service-requirements and a service-rate function $\mu(n_2)= \lambda(K_1-n_2)$, where n_2 is the number of customers at the second queue.

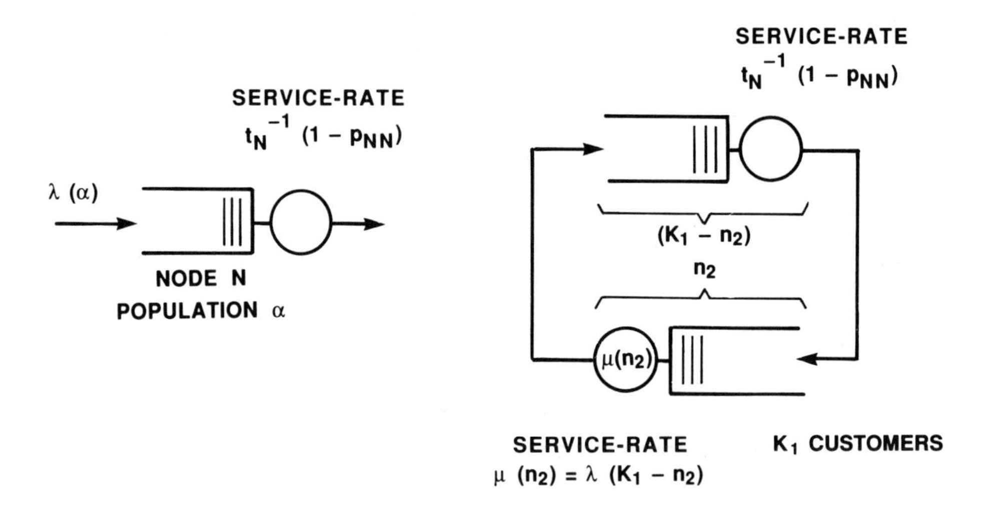

Figure 3.3
Equivalent Representations of the Reduced System **A** Arising in the Decomposition by Service Center Method

The single-level Decomposition and Agregation by Service Center procedure, as applied to the queueing network $B(N,K_1)$, may be summarized as follows:

Single-Level Decomposition by Service Center Procedure for the Queueing Network B(N,K₁):

Step 1: For $0 \leq \alpha \leq K_1$, let $S(\alpha) = \{\mathbf{n} \mid \mathbf{n} \in S, n_N = \alpha\}$ and construct the generator $\mathbf{Q}_\alpha{}^*$ corresponding to $B^*(N-1,K_1-\alpha)$, from $\mathbf{Q}(\alpha,\alpha)$ by distributing the quantity $s_{i\alpha}$ over the non-zero elements of the ith row of $\mathbf{Q}(\alpha,\alpha)$, for $1 \leq i \leq n(\alpha)$, where

$$n(\alpha) = \binom{K_1 - \alpha + N - 2}{N-2}.$$

Solve the system $\mathbf{v}^*(\mathbf{Q}^* + \mathbf{I}) = \mathbf{v}^*$ subject to the constraint that $\mathbf{v}_\alpha{}^* \mathbf{1}^T = 1$, for $0 \leq \alpha \leq K_1$. The distribution $\mathbf{v}_\alpha{}^*$ corresponds with the distribution associated with $B^*(N-1,K_1-\alpha)$.

Step 2: Construct the reduced system $\mathbf{A}$ by computing the elements a_{ij}, for $0 \leq i,j, \leq K_1$, using eq. 3.5. Solve for $\mathbf{u} = (u_0,\ldots,u_{K_1})$, where $\mathbf{u}(\mathbf{A}+\mathbf{I}) = \mathbf{u}$ and $\mathbf{u}\mathbf{1}^T = 1$ (in the case of the Gordon-Newell network with constant-speed servers, $\mathbf{u}$ is given explicitly by eq. 3.6).

Step 3: Compute the approximation for π as $\mathbf{z} = [[u_0 \mathbf{v}_0{}^*], \ldots, [u_{K_1} \mathbf{v}_{K_1}{}^*]]$.

The single-level Decomposition and Aggregation technique described above may be extended readily into a multiple-level hierarchical decomposition since each system $B^*(N-1,K_1-\alpha)$ can itself be decomposed into the set of systems $\{B^*(N-2,x) \mid 0 \leq x \leq K_1-\alpha\}$, and so on. In general, it is seen that the system $B^*(n,x)$ may be decomposed into the set of subsystems $\{B^*(n-1,y) \mid 0 \leq y \leq x\}$. Such a multiple-level decomposition is illustrated in Fig. 3.2 by the solid and broken lines.

The above general procedure may also be extended to the case in which the service-rates at the nodes are state-dependent. In this case, however, the service-rate function $\mu_x(.)$ is dependent on n_x and the explicit expression for a_{ij} is given by

$$a_{ij} = \begin{cases} t_N^{-1}(1-p_{NN})\mu_N(i), & \text{if } j = i-1, \\ \sum_{\mathbf{n}^{(N-1)} \in S(K_1-i)} v^*(\mathbf{n}^{(N-1)},K_1-i)\left(\sum_{x=1}^{N-1} t_x^{-1} p_{xN} \delta(n_x)\mu_x(n_x)\right), & \text{if } j = i+1, \\ -\sum_{\substack{k=0 \\ k \neq i}}^{K_1} a_{ik}, & \text{if } j = i, \\ 0, & \text{otherwise,} \end{cases}$$

where

$$\delta(n_x) = \begin{cases} 1, & \text{if } n_x > 0, \\ 0, & \text{otherwise.} \end{cases}$$

As in the previous case, the reduced system $\mathbf{A}$ may be interpreted as a birth-death process. The equilibrium distribution $\mathbf{u}$, $\mathbf{u} = (u_0,\ldots,u_{K_1})$, in the case of state-dependent service-rate functions is given by

$$u_k = C^{-1} \prod_{i=1}^{k} a_{(i-1)i} t_N / \mu_N(i)(1-p_{NN}).$$

From the above developments, we see that the Decomposition and Aggregation by Service Center procedure reduces the analysis of $B(N,K_1)$ to the analysis of the set of queueing networks $\{B^*(N-1,K_1-\alpha) \mid 0 \leq \alpha \leq K_1\}$ and the reduced system **A**, the supposed advantage here being that all of these systems have a smaller state-space than that of the original system $B(N,K_1)$. The computational advantage to be gained in this application of the Decomposition and Aggregation procedure is, however, actually quite mitigated in view of the efficient computational algorithms that have now been developed for the analysis of queueing networks and in view of the fact that we know *a priori* the exact forms of the state-distributions for $B(N,K_1)$ and **A**. However, the value of the Decomposition and Aggregation procedure, here, is that it exposes the structure of the queueing network in the sense of showing how the analysis may be broken down into a set of smaller related queueing network problems.

In a later work, Courtois [COU2] extended the application of the Decomposition and Aggregation by Service Center procedure to the analysis of closed multiple-chain BCMP queueing networks, to be denoted by $B(N,\mathbf{K})$, where $\mathbf{K} = (K_1,...,K_R)$ and K_r is the population of chain r. In order that we need not concern ourselves with the ordering of customers in the queues, which involves excessive notation, we will assume in the following, without loss of generality, that in the BCMP queueing network under consideration all queues of the LCFSPR and FCFS types are replaced by queues with the PS service discipline (this may be assumed since, as can be seen from eq. 2.10, all of these service disciplines have the same marginal equilibrium state-distribution). We shall also assume, without loss of generality, that there is no class-switching (from Subsection 2.1.3 it is known that the analysis of a BCMP network with class-switching may be reduced easily to the analysis of a network with no class-switching).

In the case of multiple closed routing chains, we may partition the state-space $S^{(R)}$ into the sets of states $S(\alpha)$, where $S(\alpha) = \{\mathbf{n}^{(R)} \mid \mathbf{n}^{(R)} \in S^{(R)}, \mathbf{n}_N^{(R)} = \mathbf{k}\}$, $1 \leq \alpha \leq M$, $\mathbf{k} = (k_1,...,k_R)$, $\mathbf{0} \leq \mathbf{k} \leq \mathbf{K}$, $\mathbf{0} = (0,...,0)$,

$\mathbf{n}_N^{(R)} = (n_{N1},...,n_{NR})$, n_{ir} is the number of customers of chain r at node i, $S^{(R)}$ is the state-space of $B(N,\mathbf{K})$ and R is the number of closed routing chains in $B(N,\mathbf{K})$. The number of subsystems $\mathbf{Q}(\alpha,\alpha)$ resulting from this method of state-space partitioning is $M = \prod_{r=1}^{R} (K_r+1)$ and these may be enumerated from 1 to M in such a way that subsystem $\mathbf{Q}(\alpha,\alpha)$, $\alpha = \gamma(\mathbf{k})$, is the one corresponding to the population vector $(\mathbf{K}-\mathbf{k})$, where $\gamma(.)$ is a function which maps the elements of the set $\{\mathbf{k} \mid 0 \leq k_r \leq K_r,\ r = 1,...,R\}$ into the set of elements $\{1,...,M\}$. The subsystem $\mathbf{Q}_\alpha^*$ constructed from $\mathbf{Q}(\alpha,\alpha)$, where $\alpha = \gamma(\mathbf{k})$, is the infinitesimal generator for a system defined by $B^*(N-1,\mathbf{K}-\mathbf{k})$. Hence, the system $B(N,\mathbf{K})$ is decomposed into the set of queueing networks $\{B^*(N-1,\mathbf{K}-\mathbf{k}) \mid \mathbf{k} = (k_1,...,k_R),\ 0 \leq k_r \leq K_r,\ r = 1,...,R\}$.

As a result of the above adopted method of state-space partitioning, the state-space of the reduced system $\mathbf{A}$ is $A = \{\mathbf{k} \mid 0 \leq k_r \leq K_r,\ r = 1,...,R\}$. As before, the non-zero elements a_{ij} of $\mathbf{A}$ may be derived by considering eq. 3.4. For $\mathbf{k} \in A$, $(\mathbf{k}+\mathbf{1}_r) \in A$ and $(\mathbf{k}-\mathbf{1}_r) \in A$, where $\mathbf{1}_r$ is a unit vector pointing in the direction r, we may write

$$a_{\gamma(\mathbf{k})\gamma(\mathbf{k}-\mathbf{1}_r)} = t_{Nr}^{-1}(1-p_{NN}^{(r)})\mu_N(k)k_r / k, \tag{3.7}$$

$$a_{\gamma(\mathbf{k})\gamma(\mathbf{k}+\mathbf{1}_r)} = \sum_{\mathbf{n}^{(N-1)} \in S^{(R)}(\gamma(\mathbf{k}))} v^*(\mathbf{n}^{(N-1)},\mathbf{K}-\mathbf{k}) \sum_{i=1}^{N-1} t_{ir}^{-1} p_{iN}^{(r)} \mu_i(n_i) n_{ir} / n_i, \tag{3.8}$$

and

$$a_{\gamma(\mathbf{k})\gamma(\mathbf{k})} = - \sum_{\substack{j=1 \\ j \neq \gamma(\mathbf{k})}}^{M} a_{\gamma(\mathbf{k})j},$$

where $p_{iN}^{(r)}$ is the transition probability from node i to node N for customers of chain r, $\mu_i(.)$ is the service-rate function for node i, $k = k_1+...+k_R$, $\mathbf{n}^{(N-1)} = (\mathbf{n}_1,...,\mathbf{n}_{(N-1)})$, $\mathbf{n}_i = (n_{i1},...,n_{iR})$, $n_i = n_{i1}+...+n_{iR}$ and

$v^*(\mathbf{n}^{(N-1)},\mathbf{K}-\mathbf{k})$ is the probability of state $\mathbf{n}^{(N-1)}$ in $B^*(N-1,\mathbf{K}-\mathbf{k})$. The quantity represented by eq. 3.7 may be interpreted as the transition rate for customers of chain r leaving node N and moving towards any of the nodes 1,...,N-1, given that the population vector $\mathbf{n}_N$ of node N is $\mathbf{n}_N = \mathbf{k}$. The quantity under the right-most sum of eq. 3.8 may be interpreted as the transition rate for a customer of chain r leaving node i, $1 \leq i \leq N-1$, and moving to node N, given that the state of node i in $B(N,\mathbf{K})$ is $\mathbf{n}_i$.

In the case of the multiple-chain BCMP network, the resulting reduced system is an R-dimensional birth-death process that may be interpreted as a state-dependent single-server PS queue with R types of arriving customers and arrival rates that depend on the population vector **k** of the node. In this queue, we may interpret the mean service-requirement for customers of type r as $t_{Nr}/(1-p_{NN}^{(r)})$, the service-rate function as $\mu_N(k)$ and the arrival rate $\lambda_r(\mathbf{k})$ of chain

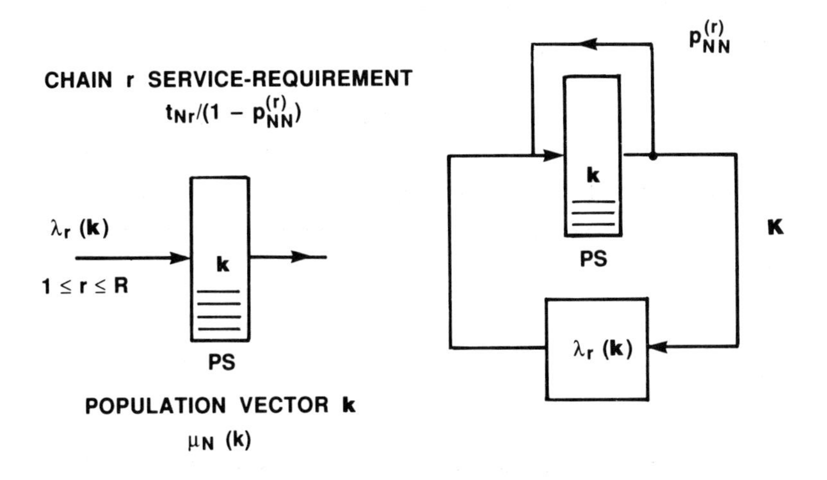

Figure 3.4
Equivalent Representations of the Reduced System **A** Assuming Multiple Types of Customers

r customers, when the population of the node is $\mathbf{k}$, as $\lambda_r(\mathbf{k}) = a_{\gamma(\mathbf{k})\gamma(\mathbf{k}+\mathbf{1}_r)}$. Alternatively, the reduced system may be interpreted as a two-queue cyclic network with R types of circulating customers and population vector $\mathbf{K}$. The first queue is one of the PS type with a return path around it with probability $p_{NN}^{(r)}$, for $1 \le r \le R$, and with service-time parameters identical to those of node N in the original network. The second queue is a special type of service-center that appropriates a total service-rate $a_{\gamma(\mathbf{k})\gamma(\mathbf{k}+\mathbf{1}_r)}$ to chain r customers when the population vector of the second queue is $(\mathbf{K}-\mathbf{k})$. At this queue, all service-requirements are assumed to be equal to one. These two equivalent interpretations are illustrated schematically in Figure 3.4.

The single-level Decomposition and Aggregation by Service Center procedure, as applied to the queueing network $B(N,\mathbf{K})$, may be summarized as follows.

Single-Level Decomposition by Service Center Procedure for the Queueing Network B(N,K):

Step 1: For $\mathbf{0} \le \mathbf{k} \le \mathbf{K}$, where $\mathbf{0} = (0,\ldots,0)$, let $S(\alpha) = \{\mathbf{n}^{(R)} \mid \mathbf{n}^{(R)} \in S^{(R)}, \mathbf{n}_N^{(R)} = \mathbf{k}\}$, where $\alpha = \gamma(\mathbf{k})$, and construct the generator $\mathbf{Q}_\alpha^*$ corresponding to $B^*(N-1,\mathbf{K}-\mathbf{k})$ from $\mathbf{Q}(\alpha,\alpha)$ by distributing the quantity $s_{i\alpha}$ over the non-zero elements of the ith row of $\mathbf{Q}(\alpha,\alpha)$ for $1 \le i \le n(\alpha)$, where

$$n(\alpha) = \prod_{r=1}^{R} \binom{K_r - k_r + N-2}{N-2}.$$

Solve the system $\mathbf{v}^*(\mathbf{Q}^*+\mathbf{I}) = \mathbf{v}^*$ subject to the constraint that $\mathbf{v}_\alpha^*\mathbf{1}^T = 1$. The distribution $\mathbf{v}_\alpha^*$ corresponds to the distribution $\mathbf{v}^*(\mathbf{n}^{(N-1)},\mathbf{K}-\mathbf{k})$ associated with $B^*(N-1,\mathbf{K}-\mathbf{k})$.

Step 2: Construct the reduced system **A** using eqs. 3.7 and 3.8 to compute the elements $a_{\gamma(\mathbf{k})\gamma(\mathbf{k}-\mathbf{1}_r)}$, $a_{\gamma(\mathbf{k})\gamma(\mathbf{k}+\mathbf{1}_r)}$ and $a_{\gamma(\mathbf{k})\gamma(\mathbf{k})}$, where $\mathbf{k} \in A$, $(\mathbf{k}+\mathbf{1}_r) \in A$ and $(\mathbf{k}-\mathbf{1}_r) \in A$. Solve for $\mathbf{u} = (u_1,\ldots,u_M)$, where $M = \prod_{r=1}^{R} (K_r+1)$, $\mathbf{u}(\mathbf{A}+\mathbf{I}) = \mathbf{u}$ and $\mathbf{u}\mathbf{1}^T = 1$.

Step 3: Compute the approximation for π as $\mathbf{z} = [[u_1\mathbf{v}_1^*], \ldots, [u_M\mathbf{v}_M^*]]$.

As for the case of the Gordon-Newell network, the single-level Decomposition and Aggregation technique for the multiple-chain BCMP queueing network may itself be extended into a multiple-level hierarchical decomposition since each system $B^*(N\text{-}1,\mathbf{K}\text{-}\mathbf{k})$, $\mathbf{0} \le \mathbf{k} \le \mathbf{K}$, may itself be decomposed into the set of queueing networks $\{B^*(N\text{-}2,\mathbf{x}) \mid \mathbf{0} \le \mathbf{x} \le (\mathbf{K}\text{-}\mathbf{k})\}$, where $\mathbf{x} = (x_1,\ldots,x_R)$, and so on.

In the case of both the Gordon-Newell network $B(N,K_1)$ and the multiple-chain BCMP network $B(N,\mathbf{K})$, it is seen that each subsystem in the multiple-level hierarchy may be broken down into the analysis of a reduced system and a set of further subsystems, where each reduced system may be characterized as either a single queue with a state-dependent arrival rate or as a closed two-queue cyclic network. As a result, the multiple-level hierarchical Decomposition by Service Center procedure reduces effectively the analysis of $B(N,K_1)$, or of $B(N,\mathbf{K})$, to that of a hierarchy of single queues or two-queue cyclic networks.

As has been mentioned, the procedures described above yield, in general, only approximate results. Theorem 3.1, however, provides the necessary and sufficient conditions under which it is possible in fact to obtain exact results. Applying the result of Theorem 3.1 to the general case of the multiple-chain BCMP queueing network, it is seen that exact results will be obtained if, and only if, $v^*(\mathbf{n}^{(N\text{-}1)},\mathbf{K}\text{-}\mathbf{k})$, where $\mathbf{n}^{(N\text{-}1)} \in S^{(R)}(\gamma(\mathbf{k}))$, is in fact equal to the state-distribution

$\pi(\mathbf{n}^{(R)})$ of $B(N,\mathbf{K})$ conditioned on the event that $\mathbf{n}_N = \mathbf{k}$. From eq. 2.10, we have

$$\pi(\mathbf{n}^{(R)}) = G_1^{-1} \prod_{i=1}^{N} f_i(\mathbf{n}_i^{(R)})$$

so that, for $\mathbf{n}^{(R)} \in S^{(R)}$,

$$\pi(\mathbf{n}^{(R)} \mid \mathbf{n}_N = \mathbf{k}) = \begin{cases} C^{-1}G_1^{-1} f_N(\mathbf{k}) \prod_{i=1}^{N-1} f_i(\mathbf{n}_i^{(R)}), \text{ if } \mathbf{n}_N = \mathbf{k}, \\ 0, \text{ otherwise,} \end{cases} \quad (3.9)$$

where C is a normalization constant. We see, however, that $(C^{-1}G_1^{-1} f_N(\mathbf{k}))$ is simply a constant. We may also see that eq. 3.9 defines exactly the state-distribution for a queueing network defined by $B(N\text{-}1,\mathbf{K}\text{-}\mathbf{k})$ that has service-time requirements t_{ir}, visit-ratios e_{ir} and service-rate functions $\mu_i(.)$, where $1 \le i \le (N\text{-}1)$ and $1 \le r \le R$. Hence, in the construction of $\mathbf{Q}_\alpha^*$ from $\mathbf{Q}(\alpha,\alpha)$, assuming that we choose not to perturb the parameters t_{ir} and $\mu_i(.)$ appearing in $\mathbf{Q}(\alpha,\alpha)$, the only other parameters that may be modified are the routing probabilities $p_{ij}^{(r)}$. Since the routing matrices $\mathbf{P}^{(r)}$, $1 \le r \le R$, directly determine the visit-ratios, the necessary and sufficient condition for exact results to be obtained by the Decomposition and Aggregation procedure, is that we construct the routing probabilities $p_{ij}^{(r)*}$, $1 \le i,j \le (N\text{-}1)$, associated with $B^*(N\text{-}1,\mathbf{K}\text{-}\mathbf{k})$ in such a way that $(e_{1r}^*,\ldots,e_{(N-1)r}^*) = c_r(e_{1r},\ldots,e_{(N-1)r})$, for $1 \le r \le R$, where c_r is an arbitrary positive constant, $\mathbf{e}_r^* = \mathbf{e}_r^*\mathbf{P}^{(r)*}$, $\mathbf{e}_r^* = (e_{1r}^*,\ldots,e_{(N-1)r}^*)$, $\mathbf{P}^{(r)*} = [p_{ij}^{(r)*}]$, and $(e_{1r},\ldots,e_{(N-1)r})$ is the vector of visit-ratios for chain r associated with the first (N-1) nodes in $B(N,\mathbf{K})$. In other words, assuming that in $B^*(N\text{-}1,\mathbf{K}\text{-}\mathbf{k})$ the service-time requirements and service-rate functions are the same as those in $B(N,\mathbf{K})$, exact results are obtained if we construct the routing matrices for $B^*(N\text{-}1,\mathbf{K}\text{-}\mathbf{k})$ in such a way that the resulting visit-ratios are the same, up to a multiplicative constant, as those in B(N,**K**). Under these conditions, the vector $(e_{1r}^*,\ldots,e_{(N-1)r}^*)$ is said to be *subparallel* [VAN1] to the vector $(e_{1r},\ldots,e_{Nr})$. The identification of these necessary and sufficient conditions for exact results to be obtained was originally made by Vantilborgh [VAN1] within the context of the Gordon-Newell network (R=1). The extension of these conditions to the case of multiple-chain BCMP queueing networks may be attributed to Courtois [COU2].

Assuming that the conditions for exact results to be obtained are satisfied, in the case of the multiple-chain BCMP network B(N,**K**), the quantity $a_{\gamma(\mathbf{k})\gamma(\mathbf{k}-\mathbf{1}_r)}$ may be interpreted simply as the *exact* conditional probability flux for chain r customers moving out of node N into the set of nodes {1,...,(N-1)}, given that $\mathbf{n}_N = \mathbf{k}$ in B(N,**K**). Under the same conditions, the quantity $a_{\gamma(\mathbf{k})\gamma(\mathbf{k}+\mathbf{1}_r)}$ may be interpreted as the exact conditional probability flux for chain r customers moving into node N from the set of nodes {1,...,(N-1)}, given that $\mathbf{n}_N = \mathbf{k}$.

3.3 Parametric Analysis and Norton's Theorem

A useful consequence of the Decomposition and Aggregation procedure is that it shows one how to construct a reduced system around a subsystem in a queueing network that may be of particular interest. We may then vary the parameters associated with this reduced system and observe the effect, without having to repetitively analyze the entire larger system. Such parametric analysis techniques are useful in certain types of performance evaluation studies.

Consider Fig. 3.1 and suppose that **Q** has been partitioned in such a way that the parameters, with respect to which we wish to carry out a parametric analysis, are contained exclusively in the off-diagonal blocks $\mathbf{Q}(\alpha,\beta)$, where $1 \le \alpha,\beta \le M$ and $\alpha \ne \beta$. Then, as can be seen from the general Simon and Ando Decomposition and Aggregation procedure, having obtained $\mathbf{v}_\alpha^*$ in Step 1, for $1 \le \alpha \le M$, each time we change the parameters of interest, we need only repeat Steps 2 and 3 since these parameters only enter directly into the construction of the reduced system **A**.

If we consider the Decomposition by Service Center approach, as has been described in Section 3.2, we see that the parameters associated with node N in the multiple-chain network $B(N,\mathbf{K})$ only enter into the analysis from Step 2 onwards. Hence, each time we vary the parameters associated with node N, that is, the mean service-requirements at node N or the service-rate function $\mu_N(.)$, we need only repeat the analysis as far as the construction and solution of **A** are concerned. We are assured that exact results will be obtained in a parametric analysis if we ensure that, in the construction of the subsystems $B^*(N-1,\mathbf{K}-\mathbf{k})$, the visit-ratios of the routing chains at the (N-1) nodes are subparallel to those associated with $B(N,\mathbf{K})$ so that the conditions of Theorem 3.1 will be fulfilled and the state-distribution associated with $B^*(N-1,\mathbf{K}-\mathbf{k})$ will be identical to that associated with $B(N-1,\mathbf{K}-\mathbf{k})$.

For simplicity, consider initially the application of a parametric analysis to $B(N,K_1)$. One particular way to construct the subsystem $B^*(N-1,K_1-k)$ is to consider an auxiliary system $B'(N,K_1-k)$ which is the same as $B(N,K_1-k)$ but with t_N replaced by 0, so that when a customer arrives at node N it is immediately injected into node x, $1 \leq x \leq N$, with probability p_{Nx}. In this situation, node N is said to be '*short-circuited*'. This queueing network is entirely equivalent, as far as the state-distribution is concerned, to one in which node N is removed from the network and in which the routing matrix $\mathbf{P} = [p_{ij}]$, of order N, is reduced by a process of deflation to a matrix $\mathbf{P}' = [p_{ij}']$, of order (N-1), where

$$p_{ij}' = p_{ij} + p_{iN}(\sum_{k=0}^{\infty} p_{NN}^k)p_{Nj},$$

or equivalently, assuming that $p_{NN} < 1$,

$p_{ij}' = p_{ij} + p_{iN}p_{Nj}/(1-p_{NN})$.

Furthermore, the visit-ratio vector $(e_1',...,e_{N-1}')$ associated with $\mathbf{P}'$ is, according to this deflation procedure, guaranteed to be subparallel to the vector $(e_1,...,e_N)$. Hence, the state-distribution $\mathbf{v}^*(\mathbf{n}^{(N-1)},K_1-k)$ for $B'(N,K_1-k)$ is identical to that associated with $B(N-1,K_1-k)$. Consequently, the quantity

$$\sum_{\substack{\mathbf{n}^{(N-1)} \in S(K_1-k) \\ n_x > 0}} v^*(\mathbf{n}^{(N-1)},K_1-k)$$

appearing in eq. 3.5 becomes the exact utilization $U_x(K_1-k)$ of the server at node x in $B'(N,K_1-k)$ and $\lambda(k)$, where

$$\lambda(k) = \sum_{x=1}^{N-1} t_x^{-1} p_{xN} U_x^*(K_1-k),$$

becomes the exact throughput at the short-circuit in the system $B'(N,K_1-k)$. One particular way to obtain the quantities $a_{k(k+1)}$ appearing in eq. 3.5 is, therefore, to analyze the special system $B'(N,K_1-k)$ for the throughput $T_N'(K_1-k)$ at the short-circuit as a function of k. This procedure of obtaining the parameters of the reduced system **A** is, however, exactly the one which is followed in the construction of a reduced network around node N using Norton's Theorem [CHA1] for queueing networks, as will be described below.

The so-called *Norton's Theorem* for queueing networks is a particular procedure for constructing an exact reduced system around an arbitrary set of nodes σ in a product-form queueing network so as to facilitate a parametric analysis with respect to the parameters that are associated with σ. The original theorem, due to Chandy et al [CHA1], presumes a Gordon-Newell type network with single entry and exit points for the set of nodes σ. The generalization for multiple-chain BCMP queueing networks and an arbitrary set σ

may be attributed to Balsamo and Iazeolla [BALS2] and Kritzinger et al [KRI2]. In the following, we first describe the procedure, as it applies to a Gordon-Newell network for the case where $\sigma = \{N\}$.

Norton's Theorem for Queueing Networks: (with $\sigma = \{N\}$)

Step 1: For $0 \leq k < K_1$: Find the throughput $T_N'(K_1\text{-}k)$ through node N in the system $B'(N,K_1\text{-}k)$ (by definition, $T_N'(0) = 0$).

Step 2: For each choice of t_N and $\mu_N(.)$: Construct the reduced system around node N, as illustrated in Figure 3.5, by replacing the set of nodes $\{1,...,(N\text{-}1)\}$ by a single-server *flow-equivalent* queue having mean service-requirements of unity and the state-dependent service-rate function $\beta(x) = T_N'(x)$. The equilibrium distribution for this reduced system is $\mathbf{u} = (u_0,...,u_{K_1})$, where

$$u_k = C^{-1} \prod_{x=1}^{k} t_N \beta(K_1\text{-}x+1)/\mu_N(x)(1\text{-}p_{NN}),$$

u_k is the probability that there are k customers at node N and C is a normalization constant.

The equilibrium distribution $\mathbf{u}$ for the reduced system, as obtained by this procedure, corresponds exactly with the true marginal equilibrium distribution for node N in $B(N,K_1)$. As has been mentioned above, this procedure for constructing an exact reduced system around node N is just one particular way of constructing an equivalent reduced system following the general Decomposition and Aggregation methodology. Having determined $\mathbf{u}$, the mean performance measures for node N may be determined readily. The mean number of customers Q_N at node N and the throughput T_N out of node N towards the set of nodes $\{1,...,(N\text{-}1)\}$ are given by

$$Q_N = \sum_{k=1}^{K_1} k u_k$$

and

$$T_N = \sum_{k=1}^{K_1} u_k t_N^{-1} \mu_N(k)(1-p_{NN}),$$

respectively. We also have, using Little's result [LIT1], that $W_N = Q_N/T_N$, where W_N is the mean waiting-time (including the service-time) for customers at node N. If in the analysis of $B'(N,K_1-k)$ (Step 1) we have also obtained the mean queue-lengths $Q_n'(K_1-k)$ and the throughputs $T_n'(K_1-k)$, for $1 \le n \le (N-1)$ and $0 \le k \le K_1$, then these may be used to obtain the mean performance measures for nodes $\{1,...,(N-1)\}$ in $B(N,K_1)$. For $1 \le n \le (N-1)$, one may write

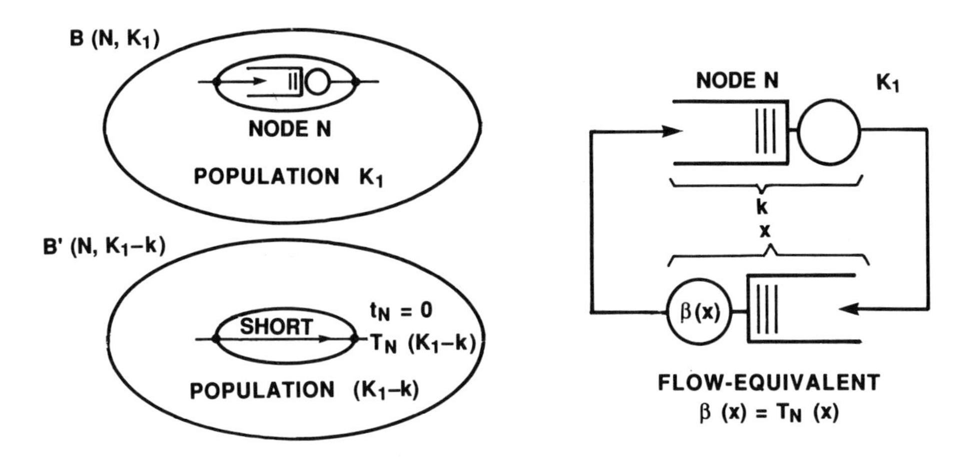

Figure 3.5
Construction of Reduced System using Norton's Theorem

$$Q_n = \sum_{k=0}^{K_1-1} u_k Q_n'(K_1-k),$$

$$T_n = \sum_{k=0}^{K_1-1} u_k T_n'(K_1-k)$$

and $W_n = Q_n / T_n$.

Although the above summarized parametric analysis procedure is given explicitly in terms of the parameters t_N and $\mu_N(.)$, we may also conduct a parametric analysis with respect to the visit-ratio e_N. This may be done in a simple way since, as can be seen from eq. 2.10, the parameter $e_N = e_{N1}$ only enters into the state-distribution $\pi(\mathbf{n}^{(1)})$ for $B(N,K_1)$ through the relative traffic intensity $w_{N1} = e_{N1}t_{N1}$, where $t_{N1} = t_N$. Hence, if we are to vary e_{N1} by a certain multiplicative factor, then we may achieve the same end effect by varying t_{N1} by the same factor.

The above developments for $\sigma = \{N\}$ may be extended to the case of the closed multiple-chain BCMP queueing network. In this situation, in Step 1 of Norton's Theorem, node N is short-circuited and the throughput $T_{Nr}'(\mathbf{k})$ of chain r customers, $1 \le r \le R$, at node N is found for the system $B'(N,\mathbf{k})$, as a function of $\mathbf{k}$, for $\mathbf{0} < \mathbf{k} \le \mathbf{K}$ and $1 \le r \le R$. In Step 2, the reduced system is constructed around node N by replacing the set of nodes $\{1,...,(N-1)\}$ by a flow-equivalent queue where all customers have mean service-requirements of unity and where there is a special type of state-dependent server that accomplishes work for chain r customers at the rate $T_{Nr}'(\mathbf{k})$, when the population vector of the flow-equivalent queue is $\mathbf{k}$. The reduced system constructed in this matter is exactly equivalent to the one arising in the Decomposition and Aggregation procedure when one ensures, in the Decomposition step, that the visit-ratio vector $(e_{1r}^*,...,e_{(N-1)r}^*)$, defined on page 94, is subparallel to the visit-ratio vector $(e_{1r},...,e_{Nr})$, for $1 \le r \le R$.

In the case of σ being an arbitrary set of nodes, the *set* of nodes σ is short-circuited and the throughput $T_{\sigma r}'(\mathbf{k})$ of chain r customers, $1 \le r \le R$, through σ is found, as a function of $\mathbf{k}$, for $\mathbf{0} < \mathbf{k} \le \mathbf{K}$ and $1 \le r \le R$. A reduced system is constructed around σ, by replacing the complementary set of nodes σ_c by a flow-equivalent queue where all customers have mean service-requirements of unity and where there is a state-dependent server that accomplishes work for chain r customers at the rate $T_{\sigma r}'(\mathbf{k})$, when the population vector of the flow-equivalent is $\mathbf{k}$. In this reduced system, the transition probability $q_j^{(r)}$ from the flow-equivalent queue to node j, where $j \in \sigma$, for customers of chain r is given by [BALS2]

$$q_j^{(r)} = \sum_{i \in \sigma_c} e_{ir} p_{ij}^{(r)} \Big/ \sum_{j \in \sigma} \sum_{i \in \sigma_c} e_{ir} p_{ij}^{(r)}.$$

An equivalent reduced system may also be constructed by following the general Decomposition and Aggregation procedure. To do this, however, it is necessary that the state-space $S^{(R)}$ associated with the infinitesimal generator $\mathbf{Q}$ of the multiple-chain BCMP queueing network $B(N,\mathbf{K})$ be decomposed according to the state $\mathbf{n}_\sigma^{(R)}$ of the set of nodes σ, where $\mathbf{n}_\sigma^{(R)} = (\mathbf{n}_i^{(R)} \mid i \in \sigma)$, $\mathbf{n}_i^{(R)} = (n_{i1},\dots,n_{iR})$, $\mathbf{n}_\sigma^{(R)} \in S_\sigma^{(R)}$ and

$$S_\sigma^{(R)} = \{\mathbf{n}_\sigma^{(R)} \mid 0 \le n_{ir} \le K_r,\ i \in \sigma,\ \sum_{j \in \sigma} n_{jr} \le K_r,\ 1 \le r \le R\}.$$

In this situation, we partition the state-space $S^{(R)}$ into the sets of states $S^{(R)}(\alpha)$, $1 \le \alpha \le M$, where

$$S^{(R)}(\alpha) = \{\mathbf{n}^{(R)} \mid \mathbf{n}^{(R)} \in S^{(R)},\ \mathbf{n}_\sigma^{(R)} = \mathbf{k}_\sigma^{(R)}\},$$

$\mathbf{k}_\sigma^{(R)} = (\mathbf{k}_i^{(R)} \mid i \in \sigma)$, $\mathbf{k}_i^{(R)} = (k_{i1},\dots,k_{iR})$, $\mathbf{k}_\sigma^{(R)} \in S_\sigma^{(R)}$, $\alpha = \tau(\mathbf{k}_\sigma^{(R)})$, and $\tau(.)$ is a function which maps the elements of $S_\sigma^{(R)}$ into the set of elements $\{1,\dots,M\}$. Having adopted this method of decomposition, the subsystem $\mathbf{Q}_\alpha^*$ constructed from $\mathbf{Q}(\alpha,\alpha)$ may be characterized as the generator for a BCMP queueing network that consists of the set of nodes σ_c and contains R routing chains, where the population of chain r is $(K_r - \sum_{i \in \sigma} k_{ir})$. It can be ensured that exact results will be obtained if, in the construction of $\mathbf{Q}_\alpha{}^*$, it is ensured that, for each chain r, the visit-ratio vector for the set of nodes σ_c is subparallel to $[e_{ir} : i \in \sigma_c]$. As in the case with $\sigma = \{N\}$, the process of short-circuiting σ_c is just one particular way of ensuring that this condition of parallelism is satisfied.

3.4 Decomposition by Routing Chain

Another specialized technique of Decomposition and Aggregation that has been proposed for the analysis of queueing networks is the so-called *Decomposition by Routing Chain* approach. This technique, developed by Conway and Georganas [CON1,CON4], reduces the analysis of a closed multiple-chain BCMP queueing network to that of a multiple-level hierarchy of single-chain closed queueing network problems. In the case where there is a single-level decomposition, the analysis of a closed queueing network with R closed routing chains is broken down into the analysis of

(1) a set of queueing networks that each have (R-1) chains and

(2) the analysis of a reduced system which may be characterized as a special type of single-chain closed queueing network.

A useful consequence of adopting this approach is that a reduced system is formed which contains only one particular type of customer (or closed routing chain). A parametric analysis can then be made with respect to the service-requirements or the routing matrix of this type without having to analyze repetitively the entire multiple-chain queueing network under consideration. This so-called *Parametric Analysis by Chain* technique will be considered in Section 3.5.

One particular application which motivates the consideration of the Decomposition by Chain approach is the problem in the modeling of computer-communication networks where one may wish to analyze the performance of a single virtual-circuit of interest which is embedded in a network. One may wish to observe the effect of various routing strategies and service-requirements without having to analyze the entire network. For such a parametric analysis, it may be advantageous to construct a simplified reduced system, around the particular virtual-circuit or routing chain of interest, especially if in the original queueing network there are a large number of chains. The Decomposition by Chain approach also provides much insight into the inherent structure of multiple-chain closed queueing networks.

Consider a closed multiple-chain BCMP queueing network $B(N,\mathbf{K})$ with R closed routing chains and suppose, for the sake of simplicity, that in this network any queueing disciplines of the FCFS and LCFSPR types are replaced by the PS service discipline. That this transformation has no effect on the marginal equilibrium distribution of the network may be seen from eq. 2.10. The state-space of this multiple-chain queueing network is denoted by $S^{(R)}$, where

$$S^{(R)} = \{\mathbf{n}^{(R)} \mid n_{ir} \geq 0, \text{ for } 1 \leq i \leq N, 1 \leq r \leq R; \sum_{i=1}^{N} n_{ir} = K_r, \text{ for } 1 \leq r \leq R\}$$

and K_r is the total population of chain r customers in the network. Now, rather than decompose the state-space $S^{(R)}$ according to the state $(n_{N1},\ldots,n_{NR})$ of the Nth node as was done in the Decomposition by Service Center approach, let us decompose the state-space according to the disposition in the network of the customers of some particular routing chain, say the Rth routing chain, without loss of generality. Let a particular distribution of the customers of chain R over the nodes of the network be denoted by $\mathbf{k}_R$, where $\mathbf{k}_R = (k_{1R},\ldots,k_{NR})$ and k_{iR} is the number of customers of chain R at node i. Also let

$$S(\mathbf{k}_R) = \{\mathbf{n}^{(R)} \mid \mathbf{n}^{(R)} \in S^{(R)},\ n_{1R} = k_{1R}, ..., n_{NR} = k_{NR}\}$$

and

$$L_R = \{\mathbf{k}_R \mid k_{iR} \geq 0, \text{ for } 1 \leq i \leq N; \sum_{i=1}^{N} k_{iR} = K_R\},$$

where L_R is the state-space of the K_R customers of chain R. Also suppose that the elements of L_R are enumerated from 1 to M by the indice α in such a way that $\alpha = \zeta(\mathbf{k}_R)$, where $\zeta(.)$ is a function that maps the elements of L_R into the set of elements $\{1,...,M\}$. Clearly,

$$\bigcup_{\mathbf{k}_R \in L_R} S(\mathbf{k}_R) = S^{(R)}.$$

If we now consider the application of the general Simon and Ando Decomposition and Aggregation procedure to the multiple-chain BCMP queueing network under consideration and partition the state-space of $\mathbf{Q}$ according to the state $\mathbf{k}_R$ of the chain R customers in the network, as illustrated in Figure 3.6, then the number M of diagonal blocks $\mathbf{Q}(\alpha,\alpha)$, $1 \leq \alpha \leq M$, in the Decomposition step (Step 1) is

$$M = \binom{K_R+N-1}{N-1},$$

this being the cardinality of the set L_R. Figure 3.7 illustrates the adopted partitioning of the infinitesimal generator $\mathbf{Q}$ in the case of B(4,(1,1)) which is a closed queueing network consisting of four queues of the PS type and two closed routing chains that each contain a single customer ($K_1=1$, $K_2=1$). In Figure 3.7, the state-vector is $(n_{11},n_{21},n_{31},n_{41},n_{12}, n_{22},n_{32},n_{42})$, where n_{ir} is the number of chain r customers at node i.

If we now construct $\mathbf{Q}^* = \mathbf{diag}(\mathbf{Q}_1{}^*,...,\mathbf{Q}_M{}^*)$ by distributing the off-diagonal block row-sum $s_{i\alpha}$ over the non-zero elements of the *i*th row of $\mathbf{Q}(\alpha,\alpha)$, then the subsystem $\mathbf{Q}_\alpha{}^*$ may be interpreted as a closed multiple-chain queueing network with state-space $S^{(R-1)}$ that consists of the first (R-1) chains of B(N,**K**). In this queueing network, the customers of chain R are fixed over the nodes of the network with distribution $\mathbf{k}_R$. Since each queue is assumed to be of the PS type, the k_{iR} customers of chain R at node i may be viewed as circulating continuously around at node i. Equivalently, we may view the system $\mathbf{Q}_\alpha^*$ as one where the customers of chain R are removed from the network and where modified state-dependent service-rate functions $\beta_i(.)$ of the form

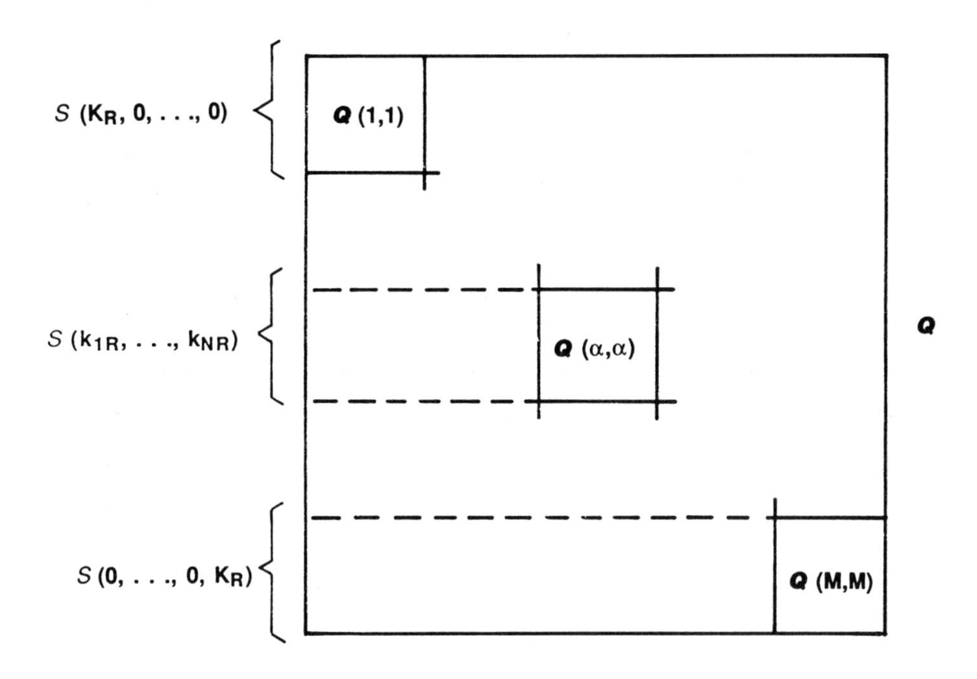

Figure 3.6
Partitioning of **Q** According to the State of Chain R Customers
(from A.E. Conway, and N.D. Georganas, Decomposition and Aggregation by Class in Closed Queueing Networks, *IEEE Trans. on Software Eng.*, 12, 10, pp. 1025-1040, 1986, with the permission of IEEE Press) © IEEE Press

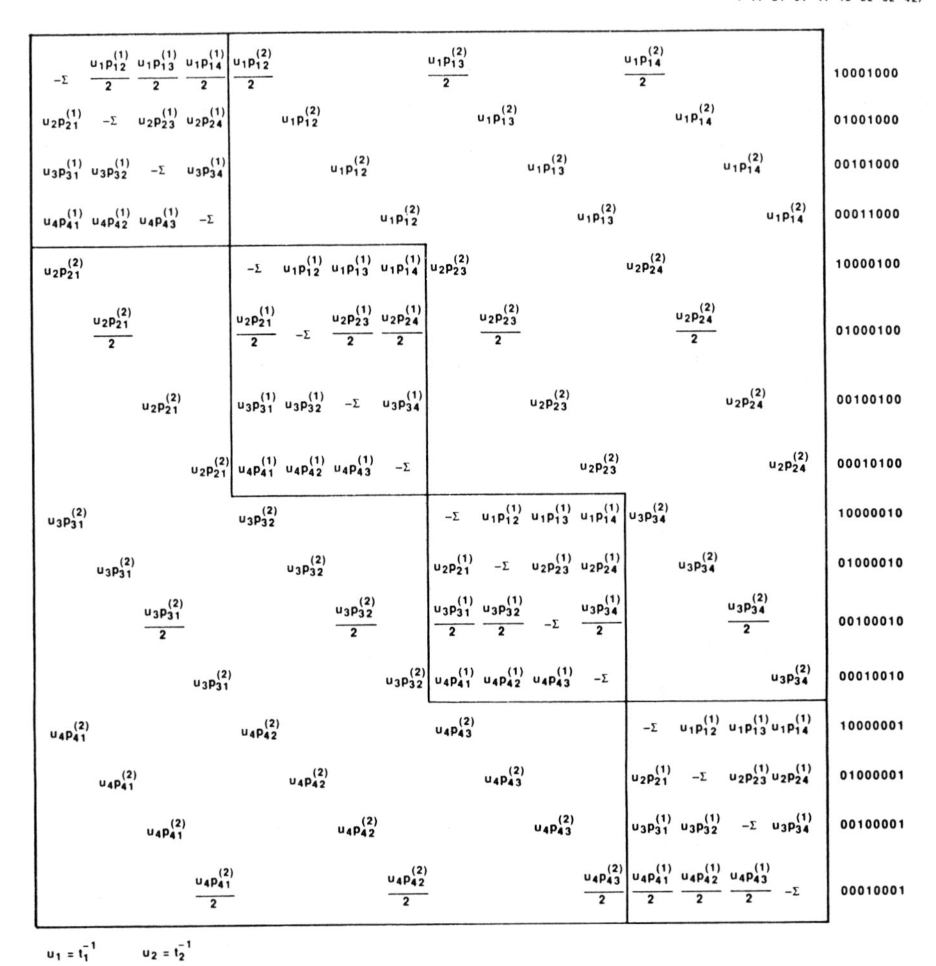

Figure 3.7
Illustration of Decomposition by Chain of the Queueing Network B(4,(1,1)) (from A.E. Conway, and N.D. Georganas, Decomposition and Aggregation by Class in Closed Queueing Networks, *IEEE Trans. on Software Eng.*, 12, 10, pp. 1025-1040, 1986, with the permission of IEEE Press) © IEEE Press

$$\beta_i(x,k_{iR}) = x\mu_i(x + k_{iR}) \,/\, (x + k_{iR}) \tag{3.10}$$

are introduced at the nodes of the network which are of the PS type. By appealing to eq. 2.10, it is seen that in both of these representations the marginal state-distribution for the customers of chains 1,...,(R-1) is identical. Yet another equivalent representation arises if we consider the k_{iR} customers of chain R at node i as k_{iR} *distinct* closed routing chains that each have a chain population of one and which loop around node i. Such routing chains have been termed '*single-customer self-looping chains*' (SCSL) in [SOU1]. These three representations for $Q_\alpha{}^*$, which are all mathematically equivalent as far as the marginal state-distribution of the customers of the first (R-1) routing chains is concerned, are illustrated schematically in Figure 3.8. We denote these equivalent queueing

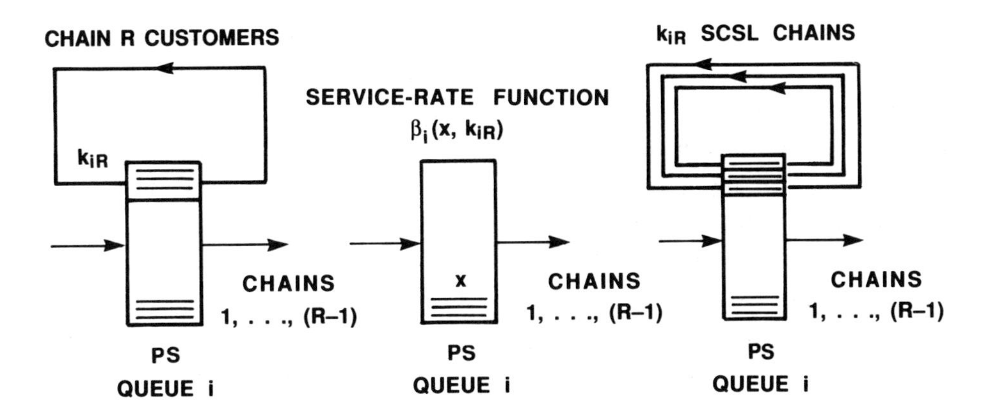

Figure 3.8
Equivalent Representations of the Queues in $M^*(R\text{-}1,\mathbf{k}_R)$

network representations for $\mathbf{Q}_{\alpha}{}^*$ as $M^*(R\text{-}1,\mathbf{k}_R)$, where $\alpha = \zeta(\mathbf{k}_R)$. Hence, in the Decomposition by Chain approach, the queueing network $B(N,\mathbf{K})$ is decomposed into the set of networks

$$\{M^*(R\text{-}1,\mathbf{k}_R) \mid \mathbf{k}_R \in L_R\}.$$

Having decomposed the queueing network $B(N,\mathbf{K})$ into the set of subsystems $\{M^*(R\text{-}1,\mathbf{k}_R) \mid \mathbf{k}_R \in L_R\}$, we now consider the reduced system $\mathbf{A}$ which is constructed in the Aggregation step of the Simon and Ando procedure. The state-space of $\mathbf{A}$, where $\mathbf{A} = [a_{\alpha\beta}]$ and $1 \le \alpha,\beta \le M$, is L_R and this is identical to the state-space of a single-chain closed queueing network with N nodes and customer population K_R. With $\alpha = \zeta(\mathbf{k}_R)$ and $\beta = \zeta(\mathbf{k}_R')$, where $\mathbf{k}_R' = (k_{1R}',...,k_{NR}')$ and $\mathbf{k}_R' \in L_R$, the element $a_{\alpha\beta}$ is non-zero *only if* $\mathbf{k}_R' = \mathbf{k}_R\text{-}\mathbf{1}_i+\mathbf{1}_j$ for some $i,j \in \{1,...,N\}$, $i \neq j$, since in any state transition only one customer of chain R may move from one node to another. According to the second step of the Decomposition and Aggregation procedure, we have $a_{\alpha\beta} = \mathbf{v}_{\alpha}{}^*\mathbf{Q}(\alpha,\beta)\mathbf{1}^T$, where $\mathbf{v}_{\alpha}{}^*$ is the distribution associated with $M^*(R\text{-}1,\mathbf{k}_R)$. The quantity $a_{\alpha\beta}$ may be interpreted as an approximation to the true conditional probability flux at which customers of chain R move from node i to node j in $B(N,\mathbf{K})$, given that $n_{1R} = k_{1R}$, ..., $n_{NR} = k_{NR}$. Since this transition rate depends on $\mathbf{k}_R$, the system $\mathbf{A}$ is a special type of single-chain queueing network consisting of the chain R customers and in which the service-rate functions at the nodes depend on the state $\mathbf{k}_R$ of the chain R customers in the *network*. Queueing networks with such general service-rate functions do not, in general, support a product-form state-distribution. However, as we shall see, the service-rate functions arising in the construction of the reduced system under consideration have a special structure that does support a product-form distribution.

The single-level Decomposition and Aggregation by Chain procedure, as applied to the queueing network $B(N,\mathbf{K})$, may be summarized as follows:

Single-Level Decomposition by Chain Procedure for the Queueing Network B(N,K):

Step 1: For $\mathbf{k}_R \in L_R$, construct the generator $\mathbf{Q}_\alpha{}^*$ corresponding to $M^*(R-1,\mathbf{k}_R)$ from $\mathbf{Q}(\alpha,\alpha)$, where $\alpha = \zeta(\mathbf{k}_R)$, by distributing the quantity $s_{i\alpha}$ over the non-zero elements of the ith row of $\mathbf{Q}(\alpha,\alpha)$ for $1 \le i \le n(\alpha)$, where

$$n(\alpha) = \prod_{r=1}^{R-1} \binom{K_r+N-1}{N-1}.$$

Solve the system $\mathbf{v}^*(\mathbf{Q}^*+\mathbf{I}) = \mathbf{v}^*$ subject to the constraint that $\mathbf{v}_\alpha{}^*\mathbf{1}^T = 1$ (the distribution $\mathbf{v}_\alpha{}^*$ corresponds to the distribution associated with $M^*(R-1,\mathbf{k}_R)$).

Step 2: Construct the reduced system $\mathbf{A}$ by computing the elements a_{ij}, for $1 \le i,j \le M$, where

$$M = \binom{K_R+N-1}{N-1},$$

using eq. 3.4. Solve for $\mathbf{u} = (u_1,\ldots,u_M)$, where $\mathbf{u}(\mathbf{A}+\mathbf{I}) = \mathbf{u}$ and $\mathbf{u}\mathbf{1}^T = 1$.

Step 3: Compute the approximation for π as $\mathbf{z} = [[u_1\mathbf{v}_1{}^*], \ldots, [u_M\mathbf{v}_M{}^*]]$.

We now turn our attention to the precise manner in which the subsystems $\mathbf{Q}_\alpha{}^*$, $1 \le \alpha \le M$, should be constructed in order to ensure that exact results will be obtained by the Decomposition and Aggregation by Chain procedure. According to Theorem 3.1, this requirement will be met if the distribution $\mathbf{v}^*$ is actually equal to the true conditional distribution $G_\alpha{}^{-1}\pi^*$. Assuming state-dependent service-rate functions, the marginal equilibrium distribution for $B(N,\mathbf{K})$ is, according to eq. 2.10,

$\pi(\mathbf{n}^{(R)}) =$

$$G_1^{-1} \{ \prod_{i=1}^{P} n_i^{(R)}!(1 / \prod_{a=1}^{n_i^{(R)}} \mu_i(a)) \prod_{r=1}^{R} w_{ir}^{n_{ir}}/n_{ir}! \} \{ \prod_{i=P+1}^{N} \prod_{r=1}^{R} w_{ir}^{n_{ir}}/n_{ir}! \}, \tag{3.11}$$

where queues 1,...,P are assumed to be of the PS type and queues P+1,...,N are of the IS type. The conditional distribution, $\pi(\mathbf{n}^{(R)} \mid n_{1R} = k_{1R}, ..., n_{NR} = k_{NR})$ is, therefore,

$\pi(\mathbf{n}^{(R)} \mid n_{1R} = k_{1R}, ..., n_{NR} = k_{NR}) =$

$$G_2^{-1}(\mathbf{k}_R)G_1^{-1}\{ \prod_{i=1}^{P} (n_i^{(R-1)}+k_{iR})!(1 / \prod_{a=1}^{n_i^{(R-1)}+k_{iR}} \mu_i(a))w_{iR}^{k_{iR}}/k_{iR}! \prod_{r=1}^{R-1} w_{ir}^{n_{ir}}/n_{ir}! \} \{ \prod_{i=P+1}^{N} w_{iR}^{k_{iR}}/k_{iR}! \prod_{r=1}^{R-1} w_{ir}^{n_{ir}}/n_{ir}!\}, \tag{3.12}$$

where

$$G_2(\mathbf{k}_R) = \sum_{\mathbf{n}^{(R)} \in S(\mathbf{k}_R)} \pi(\mathbf{n}^{(R)}).$$

Equation 3.12 may be rewritten as

$\pi(\mathbf{n}^{(R)} \mid n_{1R} = k_{1R}, ..., n_{NR} = k_{NR}) =$

$$G_3^{-1}(\mathbf{k}_R)\{ \prod_{i=1}^{P} n_i^{(R-1)}!(1 / \prod_{a=1}^{n_i^{(R-1)}} \beta_i(a,k_{iR})) \prod_{r=1}^{R-1} w_{ir}^{n_{ir}}/n_{ir}! \} \{ \prod_{i=P+1}^{N} \prod_{r=1}^{R-1} w_{ir}^{n_{ir}}/n_{ir}!\}, \tag{3.13}$$

where the function $\beta_i(a,k_{iR})$ is defined by eq. 3.10 and

$$G_3^{-1}(\mathbf{k}_R) = G_2^{-1}(\mathbf{k}_R)G_1^{-1}\{ \prod_{i=1}^{P} w_{iR}^{k_{iR}}(1/\prod_{a=1}^{k_{iR}} \mu_i(a))\}\{ \prod_{i=P+1}^{N} w_{iR}^{k_{iR}}/k_{iR}!\}.$$

The subsystem $\mathbf{Q}_\alpha{}^*$ should, therefore, be constructed in such a way that $\mathbf{v}_\alpha{}^*$ is given by eq. 3.13, but eq. 3.13 expresses precisely the state-distribution for a queueing network that contains the first (R-1) chains of B(N,**K**) and in which there are state-dependent service-rate functions $\beta_i(a,k_{iR})$ at the nodes, as given by eq. 3.10. Hence, one particular way to obtain exact results is to construct $\mathbf{Q}_\alpha{}^*$ by simply adding the off-diagonal block row-sum $s_{i\alpha}$ to the diagonal element of $\mathbf{Q}(\alpha,\alpha)$. In other words, the subsystem $\mathbf{Q}_\alpha{}^*$ should be constructed by simply *truncating* $\mathbf{Q}$ to $\mathbf{Q}(\alpha,\alpha)$. The resulting subsystem $\mathbf{Q}_\alpha{}^*$ then has eq. 3.13 as its marginal equilibrium distribution, as is required for exact results to be obtained by the general Decomposition and Aggregation procedure. We denote the subsystem $\mathbf{Q}_\alpha{}^*$ constructed in this manner by $M(R\text{-}1,\mathbf{k}_R)$.

In the case that the subsystems are constructed in the manner described above so that exact results will be obtained, the quantity $a_{\alpha\beta}$, where $\alpha = \zeta(\mathbf{k}_R)$, $\beta = \zeta(\mathbf{k}_R')$ and $\mathbf{k}_R' = \mathbf{k}_R\text{-}\mathbf{1}_i\text{+}\mathbf{1}_j$, becomes the *true* conditional probability flux at which customers of chain R move from node i to node j in B(N,**K**), given that $\mathbf{n}_R = \mathbf{k}_R$. We may write

$$a_{\alpha\beta} = \sum_{\mathbf{n}^{(R-1)} \in S^{(R-1)}} v_M(\mathbf{n}^{(R-1)},\mathbf{k}_R) t_{iR}{}^{-1} \mu_i(n_i{}^{(R-1)}+k_{iR}) p_{ij}{}^{(R)} k_{iR}/(n_i{}^{(R-1)}+k_{iR}),$$

where $v_M(\mathbf{n}^{(R-1)},\mathbf{k}_R)$ is the probability of state $\mathbf{n}^{(R-1)}$ in $M(R\text{-}1,\mathbf{k}_R)$ and $n_i{}^{(R-1)} = \sum_{r=1}^{R-1} n_{ir}$. Hence,

$$a_{\alpha\beta} = \sum_{x=0}^{K_1+\ldots+K_{R-1}} \pi_i{}^{(R-1)}(x,\mathbf{k}_R) t_{iR}{}^{-1} \mu_i(x+k_{iR}) p_{ij}{}^{(R)} k_{iR}/(x+k_{iR}), \qquad (3.14)$$

where $\pi_i{}^{(R-1)}(x,\mathbf{k}_R)$ is the marginal probability that $n_i{}^{(R-1)} = x$ in $M(R\text{-}1,\mathbf{k}_R)$.

An expression for the equilibrium distribution $\mathbf{u} = (u_1,\ldots,u_M)$ of the reduced system **A** may also be ascertained given eqs. 3.11 and 3.13. By definition,

$$u_\alpha = \pi(\mathbf{n}^{(R)}) \,/\, \pi(\,\mathbf{n}^{(R)} \mid n_{1R} = k_{1R}, \ldots, n_{NR} = k_{NR}), \tag{3.15}$$

where $\alpha = \zeta(\mathbf{k}_R)$. Furthermore, $u_\alpha = G_2(\mathbf{k}_R)$. Hence, by substitution of eqs. 3.11 and 3.12 into eq. 3.15, we obtain

$$u_\alpha = G_1^{-1}G_3(\mathbf{k}_R)\{\prod_{i=1}^{P} w_{iR}{}^{k_{iR}}(1/\prod_{a=1}^{k_{iR}}\mu_i(a))\}\{\prod_{i=P+1}^{N} w_{iR}{}^{k_{iR}}/k_{iR}!\}. \tag{3.16}$$

It is seen, therefore, that the state-distribution for the reduced system may be written in a product-form. It is not one that is contained as part of the general results of Baskett et al, Kelly or others, since it contains a term that depends on $\mathbf{k}_R$.

The single-level Decomposition and Aggregation procedure, described above, may be extended quite naturally into a hierarchical multiple-level procedure. Each of the subsystems $M(R\text{-}1,\mathbf{k}_R)$ arising in the exact single-level procedure may themselves be decomposed according to the Decomposition by Chain procedure. The system $M(R\text{-}1,\mathbf{k}_R)$, $\mathbf{k}_R \in L_R$, may be decomposed into the set of further subsystems

$$\{M(R\text{-}2,(\mathbf{k}_R+\mathbf{k}_{R-1})) \mid \mathbf{k}_{R-1} \in L_{R-1}\},$$

where L_{R-1} is the state-space of the K_{R-1} customers of chain (R-1) in $B(N,\mathbf{K})$ and $\mathbf{k}_{R-1} = (k_{1(R-1)},\ldots,k_{N(R-1)})$. The queueing network $M(R\text{-}2,(\mathbf{k}_R+\mathbf{k}_{R-1}))$ may be interpreted as one that consists of the first (R-2) chains of $B(N,\mathbf{K})$ and in which there are state-dependent service-rate functions of the form

$$\gamma_i(x,k_{i(R-1)}) = x\beta_i(x+k_{i(R-1)}+k_{iR}) \,/\, (x+k_{i(R-1)}), \tag{3.17}$$

where the function $\beta_i(.)$ is defined by eq. 3.10. Substituting eq. 3.10 into eq. 3.17, we obtain

$$\gamma_i(x,k_{i(R-1)}) = x\mu_i(x+k_{iR}+k_{i(R-1)}) \;/\; (x+k_{iR}+k_{i(R-1)}), \tag{3.18}$$

so that

$$\gamma_i(x,k_{i(R-1)}) = \beta_i(x,k_{iR}+k_{i(R-1)}).$$

Hence, it is seen that the general form of the service-rate functions for $M(R\text{-}2,(\mathbf{k}_R+\mathbf{k}_{R-1}))$, as given by eq. 3.18, is the same as that for $M(R\text{-}1,\mathbf{k}_R)$, as given by eq. 3.10.

Consider now the reduced system **A** arising in the analysis of $M(R\text{-}1,\mathbf{k}_R)$. The state of the reduced system arising in the second level of the decomposition is $\mathbf{k}_{R-1}$ and the state-space is L_{R-1}. With $\alpha = \zeta^{(2)}(\mathbf{k}_{R-1})$ and $\beta = \zeta^{(2)}(\mathbf{k}_{R-1}')$, where $\mathbf{k}_{R-1}' = \mathbf{k}_{R-1}-\mathbf{1}_i+\mathbf{1}_j$, $\mathbf{k}_{R-1}' \in L_{R-1}$ and $\zeta^{(2)}(.)$ is a function which maps the elements of L_{R-1} into the set of elements $\{1,...,M_2\}$, where

$$M_2 = \binom{K_{R-1}+N-1}{N-1}.$$

The element $a_{\alpha\beta}$ may be interpreted as the exact conditional probability flux at which chain (R-1) customers move from node i to node j in $B(N,\mathbf{K})$, given that $\mathbf{n}_R = \mathbf{k}_R$ *and* $\mathbf{n}_{R-1} = \mathbf{k}_{R-1}$. It may also be interpreted as the exact conditional probability flux at which chain (R-1) customers move from node i to node j in $M(R\text{-}1,\mathbf{k}_R)$, given that $\mathbf{n}_{R-1} = \mathbf{k}_{R-1}$. The state-distribution $\mathbf{u} = (u_1,\ldots,u_{M_2})$ of the reduced system arising in the analysis of $M(R\text{-}1,\mathbf{k}_{R-1})$ is

$$u_\alpha = G_3(\mathbf{k}_R)^{-1}G_4(\mathbf{k}_R+\mathbf{k}_{R-1})\Big\{\prod_{i=1}^{P} w_{i(R-1)}{}^{k_{i(R-1)}}\Big(1/\prod_{a=1}^{k_{i(R-1)}}\beta_i(a,k_{iR})\Big)\Big\}$$
$$\Big\{\prod_{i=P+1}^{N} w_{i(R-1)}{}^{k_{i(R-1)}}/k_{i(R-1)}!\Big\}, \tag{3.19}$$

where the function $\beta_i(.)$ is given by eq. 3.10 and $G_4(\mathbf{k}_R+\mathbf{k}_{R-1})$ is the normalization constant associated with the system $M(R\text{-}2,(\mathbf{k}_R+\mathbf{k}_{R-1}))$. Explicitly,

$$G_4(\mathbf{k}_R+\mathbf{k}_{R-1}) = \sum_{\mathbf{n}^{(R-2)} \in S^{(R-2)}} \{\prod_{i=1}^{P} n_i^{(R-2)}!\,(1/\prod_{a=1}^{n_i^{(R-2)}} \beta_i(a,k_{iR}+k_{i(R-1)}))\prod_{r=1}^{R-2} w_{ir}^{n_{ir}}/n_{ir}!\}\{\prod_{i=P+1}^{N}\prod_{r=1}^{R-2} w_{ir}^{n_{ir}}/n_{ir}!\}.$$

Comparing eq. 3.19 with eq. 3.16 and noting that $\mu_i(a) = \beta_i(a,0)$, we see that the general forms of the distribution of the reduced systems arising in the first and second levels of the analysis are the same.

If we continue to apply the Decomposition and Aggregation by Chain procedure to the systems $M(R\text{-}2,(\mathbf{k}_R+\mathbf{k}_{R-1}))$ arising in the second level, and so on, we effectively reduce the analysis of $B(N,\mathbf{K})$ into that of a hierarchy of single-chain queueing networks. At the first level, we have a single-chain queueing network that contains the K_R customers of chain R. At the second level, we have a set of single-chain queueing networks, indexed by the parameter vector $\mathbf{k}_R$, $\mathbf{k}_R \in L_R$, which contain the customers of chain (R-1). In general, at the *r*th level, $r > 1$, we have single-chain networks, indexed by $(\mathbf{k}_R+...+\mathbf{k}_{R-r+2})$, $\mathbf{k}_R \in L_R$, ..., $\mathbf{k}_{R-r+2} \in L_{R-r+2}$, which contain the customers of chain (R-r+1).

An analogy may be drawn between this hierarchical multiple-level Decomposition by Chain procedure and the hierarchical Decomposition by Service Center procedure. In the former, the multiple-chain queueing network $B(N,\mathbf{K})$ is broken down effectively into a hierarchy of single-chain queueing networks that have special state-dependent service-rate functions at the nodes. In the latter, the system $B(N,\mathbf{K})$ is broken down into a hierarchy of two-queue cyclic networks or a hierarchy of single queues with special state-dependent input arrival rates.

3.5 Parametric Analysis by Chain

One useful consequence of the single-level Decomposition and Aggregation by Chain procedure is that the reduced system **A**, constructed in the Aggregation step, may be used as a vehicle for an efficient parametric analysis with respect to a particular routing chain of interest. By analyzing the reduced system, we may determine the mean performance measures for the network as a function of the service-time requirements and the routing matrix $\mathbf{P}^{(r)}$ associated with a particular routing chain r, without having to analyze repetitively the entire larger system $B(N,\mathbf{K})$.

Consider the steps involved in the construction of the reduced system **A**, summarized as follows. We assume below that the particular routing chain of interest is the Rth, without loss of generality.

Construction of the Reduced System A in the Decomposition by Chain Method:

Step 1: For all $\mathbf{k}_R$, $\mathbf{k}_R \in L_R$:
For the queueing network $M(R\text{-}1,\mathbf{k}_R)$, determine $\pi_i^{(R-1)}(x,\mathbf{k}_R)$ for $1 \le i \le N$ and $0 \le x \le (K_1+...+K_{R-1})$.

Step 2: Construct the infinitesimal generator $\mathbf{A} = [a_{\alpha\beta}]$, where $1 \le \alpha,\beta \le M$, $M = \binom{K_R+N-1}{N-1}$, $a_{\alpha\beta}$ is given by eq. 3.14, $\alpha = \zeta(\mathbf{k}_R)$, $\beta = \zeta(\mathbf{k}_R')$, $\mathbf{k}_R \in L_R$, $\mathbf{k}_R' = \mathbf{k}_R - \mathbf{1}_i + \mathbf{1}_j$, $\mathbf{k}_R' \in L_R$ and $1 \le i,j \le N$.

Step 3: Solve the system $\mathbf{u}(\mathbf{A}+\mathbf{I}) = \mathbf{u}$ subject to the constraint that $\mathbf{u}\,\mathbf{1}^T = 1$, where $\mathbf{u} = (u_1,...,u_M)$. The marginal probability that the state of the customers of chain R is $\mathbf{k}_R$ is given by u_α, where $\alpha = \zeta(\mathbf{k}_R)$, $\mathbf{k}_R \in L_R$.

Having determined $\mathbf{u}$, the mean performance measures for chain R may be obtained readily. The throughput T_{iR} of chain R customers at node i is given by

$$T_{iR} = \sum_{\mathbf{k}_R \in L_R} u_{\zeta(\mathbf{k}_R)} \sum_{j=1}^{N} a_{\zeta(\mathbf{k}_R)\zeta(\mathbf{k}_R - \mathbf{1}_i + \mathbf{1}_j)} \delta(i,j),$$

where $a_{\zeta(\mathbf{k}_R)\zeta(\mathbf{k}_R-\mathbf{1}_i+\mathbf{1}_j)}$ is given by eq. 3.14, $\alpha = \zeta(\mathbf{k}_R)$ and $\delta(i,j) = 0$, if $i = j$, and 0 otherwise. The mean number of customers Q_{iR} of chain R at node i (including those in service) is given by

$$Q_{iR} = \sum_{\mathbf{k}_R \in L_R} k_{iR} u_{\zeta(\mathbf{k}_R)}.$$

By Little's result [LIT1], we also have

$$W_{iR} = Q_{iR} / T_{iR}.$$

Having carried out Step 1, each time we vary the parameters t_{iR}, $1 \le i \le N$, or $\mathbf{P}^{(R)} = [p_{ij}^{(R)}]$ associated with chain R, we need only repeat Steps 2 and 3 since these parameters do not enter into Step 1. Furthermore, if in Step 1 we also obtain the mean performance measures for chains 1,...,(R-1) in $M(R-1,\mathbf{k}_R)$, that is, the throughputs $T_{ir}(\mathbf{k}_R)$ and the mean queue-lengths $Q_{ir}(\mathbf{k}_R)$, for $1 \le i \le N$ and $1 \le r \le (R-1)$, then each time we vary the parameters of chain R, the measures for chains $r = 1,\ldots,R-1$ in $B(N,\mathbf{K})$ may also be obtained directly as follows:

$$T_{ir} = \sum_{\mathbf{k}_R \in L_R} T_{ir}(\mathbf{k}_R) u_{\zeta(\mathbf{k}_R)},$$

$$Q_{ir} = \sum_{\mathbf{k}_R \in L_R} Q_{ir}(\mathbf{k}_R) u_{\zeta(\mathbf{k}_R)},$$

and

$W_{ir} = Q_{ir} / T_{ir}$.

An analogy may be drawn between the Parametric Analysis by Service Center method and the Parametric Analysis by Chain method. In the former procedure, in order to construct the reduced system around a particular set of nodes σ in $B(N,\mathbf{K})$, it is required that we analyze the set of subsystems $\{B_{\sigma_c}(\mathbf{k}) \mid \mathbf{0} < \mathbf{k} \leq \mathbf{K}\}$, where $B_{\sigma_c}(\mathbf{k})$ is a multiple-chain BCMP queueing network which consists of the set of nodes σ_c and which has a population vector $\mathbf{k}$. This same set of subsystems is analyzed as part of the Norton's Theorem method. In the latter method, in order to construct the reduced system around the Rth routing chain, it is required that we analyze the set of subsystems $\{M(R\text{-}1,\mathbf{k}_R) \mid \mathbf{k}_R \in L_R\}$. In both of cases, the analysis of a set of subsystems is carried out to determine the elements $a_{\alpha\beta}$ of a reduced system $\mathbf{A}$. Hence, although these two methods of parametric analysis appear to be quite different, within the framework of the general theory of Decomposition and Aggregation we are, in fact, following the same procedure, the only fundamental difference being the particular type of state-space decomposition that is adopted initially.

3.6 Decomposition and Aggregation in Reversible Networks

In Section 2.2, we described conditions under which a queueing network of the BCMP type is reversible. An interesting theoretical consequence of reversibility in a closed BCMP queueing network is that it implies that exact results may always be obtained by the general Simon and Ando Decomposition and Aggregation procedure, *regardless* of the actual choice of state-space partitioning that is adopted in the Decomposition step. As a result, if we are interested in analyzing a closed BCMP queueing network by the method of Decomposition and Aggregation, it is advantageous if we can somehow transform it into one that is reversible. Another advantage of performing such a transformation is that it simplifies considerably the structure of the Markov chain associated with the queueing network. This then makes the inherent structure of product-form queueing networks more accessible to interpretation.

Given an arbitrary closed product-form queueing network of the BCMP type, it is possible to convert it into a reversible closed network to which it is equivalent, as far as the marginal equilibrium distribution is concerned. Consider, for the sake of simplicity, a closed multiple-chain BCMP queueing network with class-switching and state-dependent service-rate functions of the form $\mu_{ir}(n_{ir})$. If we now perform the following transformations:

(a) replace all queues of the FCFS and LCFSPR types with the PS service discipline,

(b) replace all general service-requirement distributions at PS, LCFSPR and IS queues by exponential distributions with the same means and, finally,

(c) replace the routing matrices $\mathbf{P}^{(r)} = [p_{ic;jd}^{(r)}]$, $1 \le r \le R$, by ones which satisfies the reversibility condition (eq. 2.24),

$\alpha_{ic}^{(r)} p_{ic;jd}^{(r)} = \alpha_{jd}^{(r)} p_{jd;ic}^{(r)}$, for $1 \le i,j \le N$, $c \in C_{ir}$, $d \in C_{jr}$,

where the visit-ratios $\alpha_{ic}^{(r)}$, associated with $\mathbf{P}^{(r)}$, are the same as in the original network, then it is ensured that the transformed network will be reversible. Furthermore, the same marginal equilibrium distribution, as given by eq. 2.10, is maintained since, according to the specification of $f_i(\mathbf{n}_i^{(R)})$ in eq. 2.10,

(a) the form of $f_i(.)$ for queues of the FCFS, PS and LCFSPR types is the same;

(b) the state-distribution $\pi(\mathbf{n}^{(R)})$ is insensitive to the actual forms of the service-requirement distributions at PS, LCFSPR and IS queues and, finally,

(c) the state-distribution $\pi(\mathbf{n}^{(R)})$ only depends on the routing matrices $\mathbf{P}^{(r)}$, $1 \le r \le R$, through the visit-ratios $\alpha_{ic}{}^{(r)}$.

Hence, for our purposes, there is no loss in generality in considering the reversible version of the original queueing network under consideration. Since both have the same marginal distribution, they both have the same mean performance measures and these are the quantities that are of ultimate interest here.

We now show, by means of a simple proof, that the general Simon and Ando Decomposition and Aggregation procedure always yield exact results when applied to Markov chains that satisfy the necessary and sufficient conditions for reversibility, as stated in Section 2.2.

Theorem 3.2: If the Markov chain with infinitesimal generator $\mathbf{Q}$ and state-space S is reversible and $\mathbf{Q}_\alpha{}^*$ is constructed from $\mathbf{Q}(\alpha,\alpha)$ by simply lumping the off-diagonal block row-sum $s_{i\alpha}$ onto the diagonal element of the ith row of $\mathbf{Q}(\alpha,\alpha)$, for $1 \le i \le n(\alpha)$, then the Simon and Ando Decomposition and Aggregation procedure of Section 3.1 yields exact results.

Proof: The set of states associated with $\mathbf{Q}(\alpha,\alpha)$ is $S(\alpha)$ and the true equilibrium distribution associated with $\mathbf{Q}$ is π, where $\pi = (\pi_1,\ldots,\pi_M)$ and $\pi_\alpha = (\pi_{1\alpha},\ldots,\pi_{n(\alpha)\alpha})$. Since $\mathbf{Q}$ is assumed to be reversible and $\mathbf{Q}_\alpha{}^*$ is assumed to be constructed from $\mathbf{Q}$ simply by *truncating* S to $S(\alpha)$, that is

$$\mathbf{Q}_\alpha{}^* = \mathbf{Q}(\alpha,\alpha) + \mathbf{diag}(s_\alpha),$$

where $\mathbf{s}_\alpha = (s_{1\alpha},\ldots,s_{n(\alpha)\alpha})$ and $\mathbf{diag}(\mathbf{s}_\alpha)$ is a matrix of zeros of order $n(\alpha)$ in which the ith diagonal element is replaced by $s_{i\alpha}$, the state-distribution $\mathbf{v}_\alpha{}^*$ is, according to eq. 2.19,

$$\mathbf{v}_\alpha{}^* = \pi_\alpha / \pi_\alpha \mathbf{1}^T,$$

assuming that $S(\alpha)$ is a group of communicating states when S is truncated to $S(\alpha)$, but $\pi_\alpha/\pi_\alpha\mathbf{1}^T$ is simply the true conditional distribution $G_\alpha{}^{-1}\pi_\alpha$ so that, by Theorem 3.1, it is ensured that exact results will be obtained.

It is to be noted that, in Theorem 3.2, no restrictions are placed on how we decompose the state-space S into the sets of states $\{S(\alpha) \mid 1 \leq \alpha \leq M\}$. We only assume that, in the system $\mathbf{Q}_\alpha{}^*$, the set $S(\alpha)$ is a group of communicating states under $\mathbf{Q}(\alpha,\alpha)$. This assumption, however, involves no loss of generality and is only made for the sake of convenience since, if $S(\alpha)$ is reducible into several groups of communicating states, say D groups, then $\mathbf{Q}_\alpha{}^*$ will itself be reducible into a number of subsystems and we could then replace $\mathbf{Q}_\alpha{}^*$ by the set of auxiliary subsystems, say $\mathbf{Q}^{(1)}{}_\alpha{}^*$, $\mathbf{Q}^{(2)}{}_\alpha{}^*$, ..., $\mathbf{Q}^{(D)}{}_\alpha{}^*$. We note, finally, that Theorem 3.2 may be extended readily to the case of a multiple-level Decomposition and Aggregation procedure since each subsystem $\mathbf{Q}_\alpha{}^*$, $1 \leq \alpha \leq M$, is itself reversible when $\mathbf{Q}$ is reversible.

A direct consequence of Theorem 3.2 is that, if the Decomposition and Aggregation procedure is applied to a closed *reversible* BCMP queueing network $B(N,\mathbf{K})$, then it is ensured that exact results will be obtained, regardless of the actual method of state-space partitioning that is adopted in the Decomposition step. As well, it can be ensured that exact results will be obtained in the application of a multiple-level procedure. Hence, it may no longer be surprising that exact results are obtained by the Decomposition by Service Center method of Section 3.2 and by the Decomposition by Routing Chain method of Section 3.4. Both of these methods are merely two particular ways of decomposing the system $B(N,\mathbf{K})$.

Having established Theorem 3.2, the results of Vantilborgh in [VAN1] and of Courtois in [COU2], showing that exact results may be obtained by the Decomposition by Service Center procedure, are almost immediate.

3.7 Generalized Parametric Analysis

The fact that a closed BCMP queueing network may be analyzed exactly by the Decomposition and Aggregation procedure for any particular method of state-space partitioning that we may care to adopt suggests the notion of generalized parametric analysis. Suppose that in a queueing network $B(N,\mathbf{K})$ there is some arbitrary set of parameters with respect to which we wish to carry out a parametric analysis without having to analyze repetitively the entire queueing network. In such a situation, it may be advantageous to construct a reduced system around the portion of the network that contains the parameters of interest. Since, as we have shown in Section 3.6, exact results may be obtained by the Decomposition and Aggregation procedure for any state-space partitioning, it is possible to construct such a reduced system and be assured that exact results will follow in a parametric analysis.

Consider the infinitesimal generator $\mathbf{Q}$ associated with $B(N,\mathbf{K})$ and suppose that the set of parameters of interest is Θ. If we partition the state-space $S^{(R)}$ of $\mathbf{Q}$ in such a way that the elements of Θ appear exclusively in the off-diagonal blocks $\mathbf{Q}(\alpha,\beta)$, $1 \leq \alpha,\beta \leq M$, $\alpha \neq \beta$, of $\mathbf{Q}$ and let $\mathbf{Q}_\alpha^* = \mathbf{Q}(\alpha,\alpha) + \mathbf{diag}(s_\alpha)$, then when we follow the general Simon and Ando procedure described in Section 3.1, the parameters of interest only enter into the analysis from Step 2 onwards, that is, they only enter into the reduced system $\mathbf{A}$ through the elements $a_{\alpha\beta}$, where $1 \leq \alpha,\beta \leq M$. Hence, having analyzed the set of subsystems $\{\mathbf{Q}_\alpha^* \mid 1 \leq \alpha \leq M\}$, each time we vary any of the elements of Θ, we need only repeat the Aggregation step (Step 2). This generalized method of parametric analysis includes the

Parametric Analysis by Service Center and the Parametric Analysis by Chain methods as special cases.

As a simple illustrative example, suppose we are given a multiple-chain queueing network $B(N,\mathbf{K})$ and we only wish to carry out a parametric analysis with respect to the single parameter t_{NR}. If we follow the Decomposition by Service Center method, then we would construct a reduced system around node N. As described in Section 3.2, the reduced system which is constructed with this method may be interpreted as a single queue (the Nth) with state-dependent input arrival rates $\lambda_r(.)$ that depends on the population vector $\mathbf{n}_N^{(R)}$ of the Nth node or as a two-queue cyclic network containing R types of customers. In either of these situations, each time we vary t_{NR} we have to repeat the analysis of a queueing system in which there are R types of customers. The state-space of $\mathbf{n}_N^{(R)}$ is $\{\mathbf{n}_N^{(R)} \mid 0 \leq n_{Nr} \leq K_r, 1 \leq r \leq R\}$ and therefore the cardinality of the state-space of the reduced system $\mathbf{A}$ is $\prod_{r=1}^{R} (K_r+1)$. This implies that the computational requirements involved in the analysis of $\mathbf{A}$ are exponential in R. The generalized method of parametric analysis proposed here, however, suggests a simpler parametric analysis procedure. Suppose we partition the state-space $S^{(R)}$ of $B(N,\mathbf{K})$ in such a way that

$$S(\alpha) = \{ \mathbf{n}^{(R)} \mid \mathbf{n}^{(R)} \in S^{(R)}, n_{NR} = \alpha\},$$

where $0 \leq \alpha \leq K_R$, so that t_{NR} only appears in the off-diagonal blocks $\mathbf{Q}(\alpha,\beta)$, $1 \leq \alpha,\beta \leq M$, $\alpha \neq \beta$. Then the number of subsystems M involved in the decomposition is $M = K_R+1$ and the cardinality of the state-space of $\mathbf{A}$ is (K_R+1). In this situation, each time we vary t_{NR}, we need only repeat the analysis of a Markov chain having (K_R+1) states.

The generality of the method of parametric analysis that we have proposed in this section provides a theoretical foundation for the development of new methods of exact parametric analysis that may possibly be of practical use. Such techniques may also provide the basis for new types of approximation methods for queueing networks that do not support a product-form state-distribution. These are some directions for possible future research.

CHAPTER 4

A Unified Constructive Theory for Exact Computational Algorithms

In this chapter, we formulate a general unified theory and methodology for the construction of exact computational algorithms based on the results we have established in Section 3.6. In Theorem 3.2 of Section 3.6, we established that the Simon and Ando Decomposition and Aggregation procedure always yield exact results for reversible Markov chains. We also explained how a BCMP queueing network could be transformed, without loss of generality, into one that satisfies the necessary and sufficient conditions for reversibility. As a result, the procedure may always yield exact results when applied to closed, reversible BCMP queueing networks, regardless of the actual method of state-space partitioning that is adopted in the Decomposition step. We also established we can ensure that exact results will be obtained using a multiple-level Decomposition and Aggregation procedure. These observations suggest the notion of a generalized Decomposition and Aggregation procedure for closed BCMP queueing networks, based on a hierarchy of interrelated subsystems.

In the first section of this chapter, we formalize such a generalized hierarchical Decomposition and Aggregation procedure. In Section 4.2, we then identify the possible existence of equivalent subsystems at each level in the hierarchy. Such subsystems, which are identical as far as their equilibrium state-distributions are concerned, are said to belong to the same *equivalence class*. The existence of such equivalence classes leads one to the notion of the existence of a generalized recursive structure that may be identified in a queueing network. It is this recursive structure which provides the theoretical basis for the general unified methodology for the construction of efficient computational algorithms to be formulated in Section 4.3.

Having established a general methodology for the construction of computational algorithms, we then show, in Sections 4.4 to 4.8 how the main exact computational algorithms that have been developed may be formulated in terms of this methodology. This serves to unify, in a simple way, these developments that have been made. We then consider, in Section 4.9, the possibility of hybrid algorithms based on hybrid methods of state-space partitioning. Finally, in Section 4.10, we consider the existence of more efficient algorithms than have hitherto been developed. As will be seen, the unified theory suggests, in fact, the existence of a range of new algorithms that can possibly be developed. The theory also suggests the existence of an algorithm which is computationally optimal within the class of algorithms defined by the proposed methodology.

4.1 Generalized Hierarchical Decomposition of BCMP Queueing Networks

Consider the infinitesimal generator $\mathbf{Q}$ associated with a closed multiple-chain BCMP queueing network $B(N,\mathbf{K})$ with state-space $S^{(R)}$ and suppose that the network has been transformed, according to the steps described in Section 3.6, into one that is reversible. In Section 3.6, it has been established that for any particular method of state-space partitioning we may care to adopt, it is ensured that exact results will be obtained by the Simon and Ando Decomposition and Aggregation procedure. In general then, suppose that $\mathbf{Q}$ is decomposed into the set of subsystems $\{\mathbf{Q}_{\alpha_1}{}^* \mid 1 \leq \alpha_1 \leq M\}$, where $\mathbf{Q}_{\alpha_1}{}^*$ is constructed by simply truncating $\mathbf{Q}$ to a set of states $S(\alpha_1)$ that is assumed to communicate under $\mathbf{Q}(\alpha_1,\alpha_1)$ and where

$$\bigcup_{\alpha_1=1}^{M} S(\alpha_1) = S^{(R)}.$$

Since $\mathbf{Q}$ is assumed to be reversible, the subsystems $\mathbf{Q}_{\alpha_1}{}^*$, $1 \le \alpha_1 \le M$, are themselves reversible Markov chains. Hence, each of these subsystems may itself, in turn, also be analyzed exactly by a Decomposition and Aggregation procedure. In this *second* 'level' of analysis, suppose that the particular subsystem $\mathbf{Q}_{\alpha_1}{}^*$ is decomposed into the set of subsystems $\{\mathbf{Q}_{\alpha_1\alpha_2}{}^* \mid 1 \le \alpha_2 \le \chi(\alpha_1)\}$, where $\chi(\alpha_1)$ is the total number of subsystems into which $\mathbf{Q}_{\alpha_1}{}^*$ is decomposed. Also, assume that the state-space associated with $\mathbf{Q}_{\alpha_1\alpha_2}{}^*$ is the set of communicating states $S(\alpha_1,\alpha_2)$, such that $S(\alpha_1) \supset S(\alpha_1,\alpha_2)$ and

$$\bigcup_{\alpha_2=1}^{\chi(\alpha_1)} S(\alpha_1,\alpha_2) = S(\alpha_1).$$

The set of all subsystems obtained at the *second* level is, therefore, $\{\mathbf{Q}_{\alpha_1\alpha_2}{}^* \mid 1 \le \alpha_1 \le M,\ 1 \le \alpha_2 \le \chi(\alpha_1)\}$ and each of the subsystems $\mathbf{Q}_{\alpha_1\alpha_2}{}^*$ contained in this set is also a reversible Markov chain since $\mathbf{Q}_{\alpha_1}{}^*$ is itself reversible. In general, if we extend this decomposition of $B(N,\mathbf{K})$ to a multiple-level hierarchy, the set of subsystems obtained at the *n*th level, for $n \ge 2$, is $\{\mathbf{Q}_{\alpha_1\alpha_2\cdots\alpha_n}{}^* \mid 1 \le \alpha_1 \le M,\ 1 \le \alpha_i \le \chi(\alpha_1,\alpha_2,\ldots,\alpha_{i-1}),\ i = 2,\ldots,n\}$, where $\chi(\alpha_1,\alpha_2,\ldots,\alpha_{i-1})$ is the number of subsystems into which $\mathbf{Q}_{\alpha_1\cdots\alpha_{i-1}}{}^*$ is decomposed and the state-space associated with the system $\mathbf{Q}_{\alpha_1\alpha_2\cdots\alpha_n}{}^*$ is the set of communicating states $S(\alpha_1,\alpha_2,\ldots,\alpha_n)$, $S(\alpha_1,\alpha_2,\ldots,\alpha_{n-1}) \supset S(\alpha_1,\alpha_2,\ldots,\alpha_n)$. Here, it is also assumed that

$$\bigcup_{\substack{1 \le \alpha_1 \le M \\ 1 \le \alpha_i \le \chi(\alpha_1,\ldots,\alpha_{i-1}),\ 2 \le i \le n}} S(\alpha_1,\alpha_2,\ldots,\alpha_n) = S^{(R)}.$$

In the hierarchical decomposition described above, the total number of subsystems $\Omega(n)$ obtained at the *n*th level in the hierarchy is, for $n \geq 2$,

$$\Omega(n) = \sum_{\alpha_1=1}^{M} \sum_{\alpha_2=1}^{\chi(\alpha_1)} \sum_{\alpha_3=1}^{\chi(\alpha_1,\alpha_2)} \cdots \sum_{\alpha_{n-1}=1}^{\chi(\alpha_1,\ldots,\alpha_{n-2})} \chi(\alpha_1,\ldots,\alpha_{n-1}). \qquad (4.1)$$

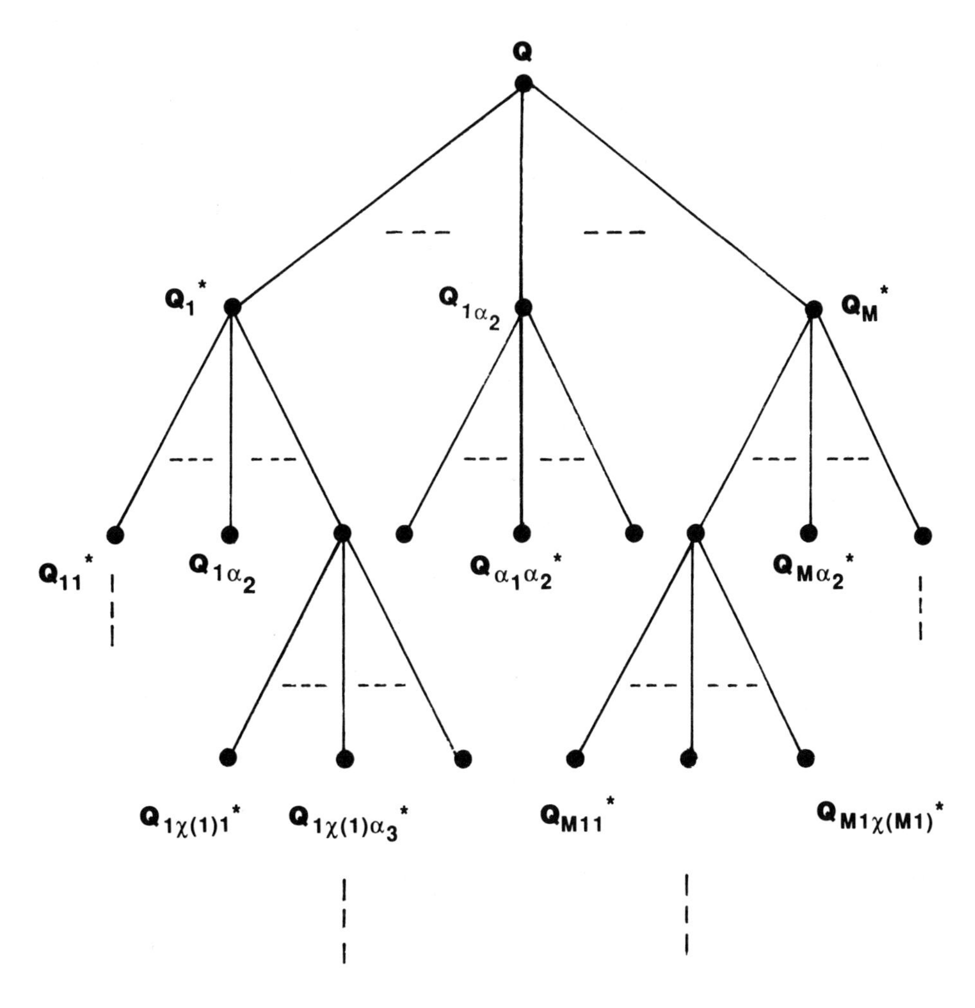

Figure 4.1
General Hierarchical Multiple-Level Decomposition of B(N,**K**)

At the first level, we have $\Omega(1) = M$. We say that $\Omega(n)$ is the *width* of level n. Figure 4.1 illustrates the hierarchical multiple-level decomposition method described above.

Having adopted some hierarchical decomposition of a closed BCMP queueing network, we may then apply the general Decomposition and Aggregation procedure, as described in Section 3.1. Assuming that there are a total of L levels in the adopted hierarchy, the hierarchical procedure consists essentially of beginning at level L and successively analyzing the subsystems at levels L-1, L-2, ..., and so on, until one has obtained the state-distribution π for $\mathbf{Q}$. The initial conditions in this hierarchical procedure are the exact state-distributions $\mathbf{v}_{\alpha_1 \cdots \alpha_L}{}^*$ associated with the subsystems $\mathbf{Q}_{\alpha_1 \cdots \alpha_L}{}^*$ at level L. Consider, for example, the analysis of a particular subsystem $\mathbf{Q}_{\alpha_1 \alpha_2 \cdots \alpha_n}{}^*$ at the *n*th level,

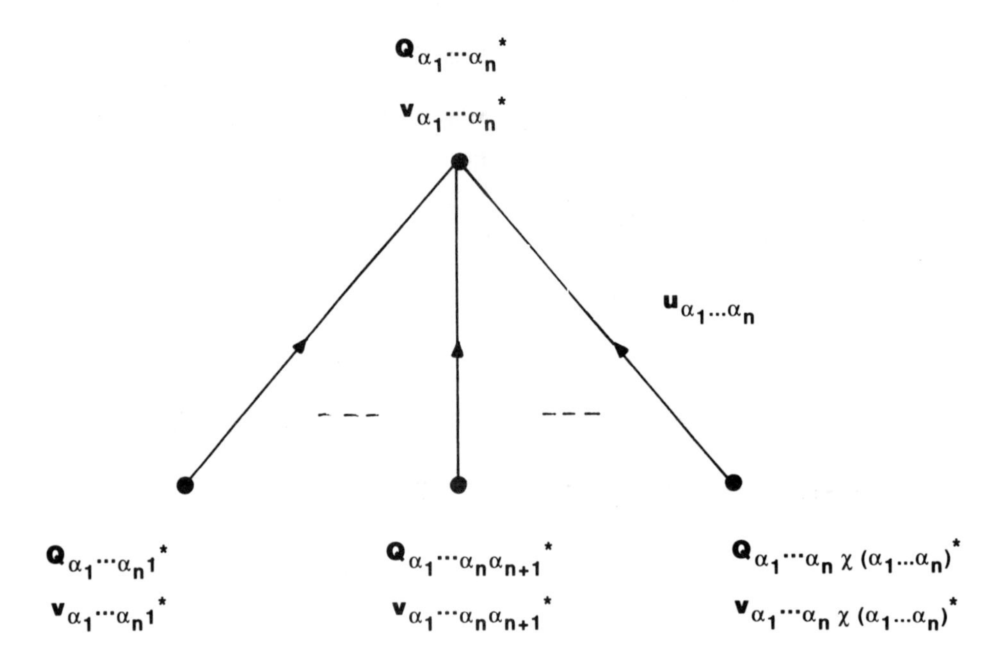

Figure 4.2
Analysis of $\mathbf{Q}_{\alpha_1 \cdots \alpha_n}{}^*$ by Decomposition and Aggregation

$1 \leq n \leq L\text{-}1$, as illustrated in Figure 4.2. Assuming that we have obtained the equilibrium state-distributions $\mathbf{v}_{\alpha_1\alpha_2\cdots\alpha_n\alpha_{n+1}}{}^*$ of the subsystems $\mathbf{Q}_{\alpha_1\alpha_2\cdots\alpha_n\alpha_{n+1}}{}^*$ for $1 \leq \alpha_{n+1} \leq \chi(\alpha_1,\alpha_2,\ldots,\alpha_n)$, we may construct the infinitesimal generator of the reduced system $\mathbf{A}_{\alpha_1\alpha_2\cdots\alpha_n}$ associated with $\mathbf{Q}_{\alpha_1\alpha_2\cdots\alpha_n}{}^*$ using the procedure described in Section 3.1. The cardinality of the state-space of this reduced system is $\chi(\alpha_1,\alpha_2,\ldots,\alpha_n)$ since this is the number of subsets of states into which $S(\alpha_1,\ldots,\alpha_n)$ is partitioned. Having obtained the state-distribution $\mathbf{u}_{\alpha_1\alpha_2\cdots\alpha_n}$ associated with the reduced system $\mathbf{A}_{\alpha_1\alpha_2\cdots\alpha_n}$, we may then obtain $\mathbf{v}_{\alpha_1\alpha_2\cdots\alpha_n}{}^*$ by deconditioning the (conditional) distributions $\mathbf{v}_{\alpha_1\alpha_2\cdots\alpha_n\alpha_{n+1}}{}^*$, $1 \leq \alpha_{n+1} \leq \chi(\alpha_1,\alpha_2,\ldots,\alpha_n)$, of the subsystems using the (marginal) distribution $\mathbf{u}_{\alpha_1\alpha_2\cdots\alpha_n}$. We may write

$$\mathbf{v}_{\alpha_1\cdots\alpha_n}{}^* = [u_1\mathbf{v}_{\alpha_1\ldots\alpha_n 1}{}^*,\ldots,u_{\chi(\alpha_1,\ldots,\alpha_n)}\mathbf{v}_{\alpha_1\ldots\alpha_n\chi(\alpha_1,\ldots,\alpha_n)}{}^*], \tag{4.2}$$

where

$$(u_1,u_2,\ldots,u_{\chi(\alpha_1,\ldots,\alpha_n)}) = \mathbf{u}_{\alpha_1\alpha_2\cdots\alpha_n}.$$

The general multiple-level hierarchical Decomposition and Aggregation procedure, as applied to the closed, reversible BCMP queueing network $B(N,\mathbf{K})$ with state-distribution π, may be summarized as follows:

General L-Level Hierarchical Decomposition and Aggregation Procedure for B(N,K):

Initialization:
Obtain the state-distributions $\mathbf{v}_{\alpha_1\alpha_2\cdots\alpha_L}{}^*$ of the subsystems $\mathbf{Q}_{\alpha_1\cdots\alpha_L}{}^*$ at level L for all $1 \leq \alpha_1 \leq M$, $1 \leq \alpha_i \leq \chi(\alpha_1,\alpha_2,\ldots,\alpha_{i-1})$, where $2 \leq i \leq L$.

Main Recursion:
For $n = (L-1),(L-2),\ldots,1$:
For $(\alpha_1,\ldots,\alpha_n) \in \{(\alpha_1,\ldots,\alpha_n) \mid 1 \leq \alpha_1 \leq M,\ 1 \leq \alpha_i \leq \chi(\alpha_1,\ldots,\alpha_{i-1}),\ i = 2,\ldots,n\}$:

Construct the reduced system $\mathbf{A}_{\alpha_1\alpha_2\cdots\alpha_n}$ using the set of distributions $\{\mathbf{v}_{\alpha_1\cdots\alpha_n\alpha_{n+1}}{}^* \mid 1 \le \alpha_{n+1} \le \chi(\alpha_1,\ldots,\alpha_n)\}$.
Find the state-distribution $\mathbf{u}_{\alpha_1\alpha_2\cdots\alpha_n}$ associated with $\mathbf{A}_{\alpha_1\alpha_2\cdots\alpha_n}$.
Find the state-distribution $\mathbf{v}_{\alpha_1\cdots\alpha_n}{}^*$ associated with the subsystem $\mathbf{Q}_{\alpha_1\cdots\alpha_n}{}^*$ using eq. 4.2.

Construct the reduced system $\mathbf{A}$ using the set of distributions $\{\mathbf{v}_{\alpha_1} \mid 1 \le \alpha_1 \le M\}$. Find the state-distribution $\mathbf{u} = (u_1,\ldots,u_M)$ associated with $\mathbf{A}$. Find π, where $\pi = [u_1\mathbf{v}_1{}^*,\ldots,u_M\mathbf{v}_M{}^*]$.

The hierarchical Decomposition and Aggregation procedure, summarized above, for closed, reversible BCMP queueing networks is, of course, extremely general and includes the Decomposition by Service Center method of Section 3.2 and the Decomposition by Routing Chain method of Section 3.4 as special cases. It is apparent, however, that if we attempt to follow this hierarchical procedure to analyze a network $B(N,\mathbf{K})$ for π, the computational costs will be excessive since the width $\Omega(n)$ of level n, as given by eq. 4.1, is in general very large. For instance, suppose that we have carried out an L-level hierarchical decomposition of $B(N,\mathbf{K})$ to the extreme extent that the cardinalities of the sets $S(\alpha_1,\ldots,\alpha_L)$ are all unity. In this situation, each subsystem $\mathbf{Q}_{\alpha_1\cdots\alpha_L}{}^*$ consists of a *single* state and the width $\Omega(L)$ of level L is equal to the cardinality of the state-space $S^{(R)}$ of $B(N,\mathbf{K})$.

As it stands, the general hierarchical method of analysis described above is, in general, an impractical method of solution. An important observation that may be made, however, is that, having adopted some particular method of hierarchical state-space partitioning, it is quite possible for a number of the subsystems in each particular level n, $2 \le n \le L$, to be equivalent as far as their state-distributions are concerned. It is also conceivable that certain subsystems in *different* levels be equivalent. Such subsystems with equivalent state-distributions may be said to belong to the same *equivalence class*. More formally, we say that two subsystems $\mathbf{Q}_{\alpha_1 \cdots \alpha_a}{}^*$ and $\mathbf{Q}_{\beta_1 \cdots \beta_b}{}^*$ belong to the same equivalence class if $\mathbf{v}_{\alpha_1 \cdots \alpha_a}{}^* = \mathbf{v}_{\beta_1 \cdots \beta_b}{}^*$. An important consequence of the possible existence of equivalent subsystems at the various levels in the hierarchy is that, having obtained the state-distribution for one of the members of an equivalence class, it is no longer necessary that we obtain the state-distribution for any of the other members. In the following section, we exploit this observation to simplify considerably the general hierarchical Decomposition and Aggregation procedure.

4.2 Equivalence Classes and Recursive Structures in Queueing Networks

The possible existence of equivalence classes in the various levels of a hierarchical decomposition leads one to the notion of the existence of a general recursive structure to be found in a queueing network. This structure may be exploited for the construction of recursive solution algorithms that are computationally efficient.

Consider the multiple-level Decomposition and Aggregation method, as described in the previous section. At the *n*th level in the hierarchy, the set of subsystems is

$$\{\mathbf{Q}_{\alpha_1 \cdots \alpha_n}{}^* \mid 1 \le \alpha_1 \le M,\ 1 \le \alpha_i \le \chi(\alpha_1, \ldots, \alpha_{i-1}),\ i = 2, \ldots, n\}.$$

As has been illustrated in Fig. 4.2, the subsystem $\mathbf{Q}_{\alpha_1 \cdots \alpha_n}{}^*$ contained in this set may be related to the set of subsystems $\{\mathbf{Q}_{\alpha_1 \cdots \alpha_n \alpha_{n+1}}{}^* \mid 1 \le \alpha_{n+1} \le \chi(\alpha_1, \ldots, \alpha_n)\}$. Hence, in the general Decomposition and Aggregegation algorithm, to obtain $\mathbf{v}_{\alpha_1 \cdots \alpha_n}{}^*$ for all subsystems at the *n*th level, one must first have obtained $\mathbf{v}_{\alpha_1 \cdots \alpha_n \alpha_{n+1}}{}^*$ for all the subsystems at level (n+1). Now consider the possible existence of equivalence classes at level n in the hierarchy. Let the set of equivalence classes at level n be denoted by $E(\mathrm{n})$ and let the equivalence classes $C(\mathrm{n})$ contained in $E(\mathrm{n})$ be enumerated from 1 to $\mathrm{E_n}$ so that $E(\mathrm{n}) = \{C_1(\mathrm{n}),\ldots,C_{\mathrm{E_n}}(\mathrm{n})\}$. By definition, let $E(0) = \mathbf{Q}$, $C_1(0) = \mathbf{Q}$ and $\mathrm{E}_0 = 1$. Also suppose that we have a function $\Phi_{\mathrm{n}}(.)$ which maps the set of subsystems $\{\mathbf{Q}_{\alpha_1\cdots\alpha_n}{}^* \mid 1 \le \alpha_1 \le \mathrm{M},\ 1 \le \alpha_i \le \chi(\alpha_1,\ldots,\alpha_{i-1}),\ i = 2,\ldots,n\}$ into the set $E(\mathrm{n})$ of equivalence classes, where $\Phi_{\mathrm{n}}(\alpha_1,\ldots,\alpha_n)$ is the particular equivalence class $C_x(\mathrm{n})$ to which $\mathbf{Q}_{\alpha_1\cdots\alpha_n}{}^*$ belongs. By the definition of an equivalence class, $\Phi_{\mathrm{n}}(\alpha_1,\ldots,\alpha_n) = \Phi_{\mathrm{n}}(\beta_1,\ldots,\beta_n)$ if $\mathbf{v}_{\alpha_1\cdots\alpha_n}{}^* = \mathbf{v}_{\beta_1\cdots\beta_n}{}^*$. Let the distribution which is common to all of the subsystems contained in $C_x(\mathrm{n})$ be denoted by $\mathbf{w}_x(\mathrm{n})$. Hence, $\mathbf{w}_x(\mathrm{n}) = \mathbf{v}_{\alpha_1\cdots\alpha_n}{}^*$ for all $1 \le \alpha_1 \le \mathrm{M}$, $1 \le \alpha_i \le \chi(\alpha_1,\ldots,\alpha_{i-1})$, $i = 2,\ldots,n$, such that $\Phi_{\mathrm{n}}(\alpha_1,\ldots,\alpha_n) = x$. This mapping of subsystems $\mathbf{Q}_{\alpha_1\cdots\alpha_n}{}^*$ into equivalence classes at level n is illustrated pictorially in Figure 4.3.

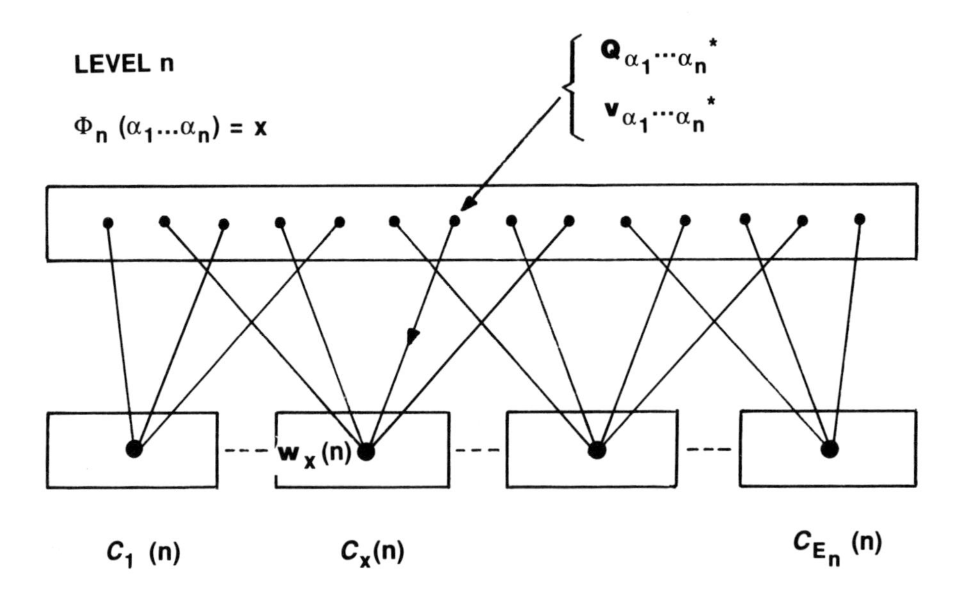

Figure 4.3
Mapping of Subsystems at Level n into Equivalence Classes

Having defined the set $E(n)$ of equivalence classes $C_x(n)$ that may be found at level n, it is straightforward to relate the state-distribution $\mathbf{w}_x(n)$ to those associated with the equivalence classes $C_x(n+1)$, $1 \le x \le E_{n+1}$, at level (n+1). Consider the determination of $\mathbf{w}_x(n)$. This is entirely equivalent to determining $\mathbf{v}_{\alpha_1 \cdots \alpha_n}{}^*$, assuming that $\mathbf{Q}_{\alpha_1 \cdots \alpha_n}{}^*$ is an element of $C_x(n)$. According to the single-level Decomposition and Aggregation procedure, to determine $\mathbf{v}_{\alpha_1 \cdots \alpha_n}{}^*$, we need $\mathbf{v}_{\alpha_1 \cdots \alpha_n \alpha_{n+1}}{}^*$ for all $1 \le \alpha_{n+1} \le \chi(\alpha_1, \ldots, \alpha_n)$. Assuming that $\Phi_{n+1}(\alpha_1, \ldots, \alpha_{n+1}) = y$, we have $\mathbf{v}_{\alpha_1 \cdots \alpha_{n+1}}{}^* = \mathbf{w}_y(n+1)$, where $\mathbf{w}_y(n+1)$ is the state-distribution at level (n+1) associated with the equivalence class $C_y(n+1)$. Hence, to obtain $\mathbf{w}_x(n)$ for $1 \le x \le E_n$, we need only have available the set of distributions $\{\mathbf{w}_y(n+1) \mid 1 \le y \le E_{n+1}\}$.

Having made the above observations, we may now define a general recursive L-level Decomposition and Aggregation procedure for reversible BCMP queueing networks that exploits the inherent redundancy due to the possible existence of equivalence classes at the various levels. This recursive procedure is illustrated in Figure 4.4 and may be summarized as follows:

General L-Level Hierarchical Decomposition and Aggregation Procedure for B(N,K) Based on Equivalence Classes:

Initialization:

For $1 \le x \le E_L$:

Obtain the state-distribution $\mathbf{w}_x(L)$ associated with the equivalence class $C_x(L)$ at level L.

Main Recursion:

For n = (L-1),(L-2),...,1:

For $1 \le x \le E_n$:

Construct the reduced system $\mathbf{A}_{\alpha_1 \cdots \alpha_n}$, where $\Phi_n(\alpha_1,\ldots,\alpha_n) = x$, using the set of state-distributions $\{\mathbf{w}_y(n+1) \mid \mathbf{w}_y(n+1) = \mathbf{v}_{\alpha_1 \cdots \alpha_n \alpha_{n+1}}{}^*,\ y = \Phi_{n+1}(\alpha_1,\ldots,\alpha_{n+1}),\ 1 \le \alpha_{n+1} \le \chi(\alpha_1,\ldots,\alpha_n)\}$.
Find the state-distribution $\mathbf{u}_{\alpha_1 \cdots \alpha_n}$ associated with $\mathbf{A}_{\alpha_1 \cdots \alpha_n}$.
Find the state-distribution $\mathbf{w}_x(n)$, where $\mathbf{w}_x(n) = \mathbf{v}_{\alpha_1 \cdots \alpha_n}{}^*$, $\Phi_n(\alpha_1,\ldots,\alpha_n) = x$ and $\mathbf{v}_{\alpha_1 \cdots \alpha_n}{}^*$ is given by eq. 4.2.

Construct the reduced system $\mathbf{A}$ using the set of distributions $\{\mathbf{v}_{\alpha_1}{}^* \mid 1 \le \alpha_1 \le M\}$, where $\mathbf{v}_{\alpha_1}{}^* = \mathbf{w}_x(1)$ and $\Phi_1(\alpha_1) = x$.
Find the state-distribution $\mathbf{u} = (u_1,\ldots,u_M)$ associated with $\mathbf{A}$.
Compute π, where $\pi = [u_1\mathbf{v}_1{}^*,\ldots,u_M\mathbf{v}_M{}^*]$.

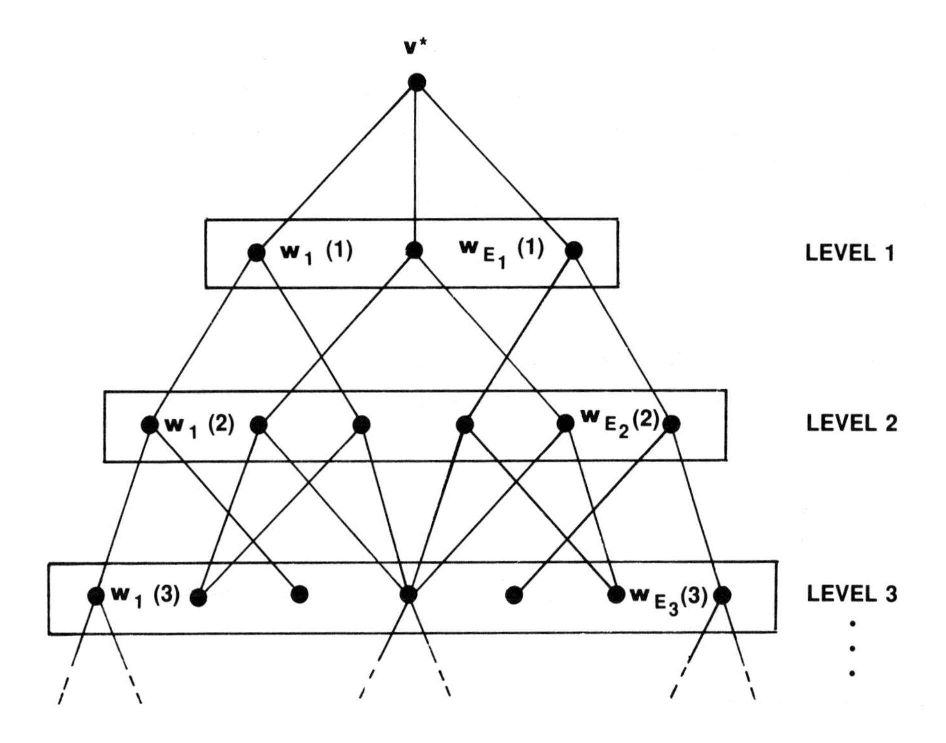

Figure 4.4
General Hierarchical Decomposition and Aggregation Procedure Based on Equivalence Classes

As is evident, the Decomposition and Aggregation procedure summarized above, based on the state-distributions of equivalence classes is, in general, more efficient than the general procedure described previously in Section 4.1. In the algorithm of Section 4.1, the width of level n is $\Omega(n)$, while in the algorithm summarized above the width is only E_n, where $E_n \leq \Omega(n)$. The width of each level and the total number of levels in a particular hierarchical decomposition are the factors that directly determine the computational requirements involved in carrying out the multiple-level analysis. If for a particular method of state-space partitioning that is adopted, it so happens that E_n is much smaller than $\Omega(n)$, then the resulting recursion will be computationally efficient. The actual efficiency obtained, however, depends completely on the actual method of state-space partitioning that is adopted since this partitioning determines the equivalence classes that may be identified. The manner of carrying out this partitioning is, of course, left completely open to us.

The general algorithms that we have formulated in this and the previous section involve recursions which are defined in terms of the equilibrium state-distributions of subsystems and the parameters and state-distributions of reduced systems. It is possible, however, that we may also be able to define the same recursive structure in terms of less abstract measures such as mean queue-lengths, throughputs, or marginal queue-length distributions since all of these measures are, by definition, simply specific functions of certain state-probabilities. In the analysis of queueing networks, it is these 'mean performance measures' which are of ultimate practical interest. In the following section, we make use of the general recursive structure that we have identified in this section to formulate a unified theory for the construction of efficient computational algorithms to obtain these mean performance measures.

4.3 A Unified Methodology for the Construction of Exact Computational Algorithms

The general hierarchical recursive structure based on equivalence classes formulated in the previous section provides the basis for a unified methodology that may be used for the construction of efficient exact computational algorithms for multiple-chain BCMP queueing networks.

Consider the equivalence class $C_x(n)$, $C_x(n) \in E(n)$, at level n in the hierarchical decomposition. As stated in the previous section, the set of subsystems in $C_x(n)$ is

$$\{\mathbf{Q}_{\alpha_1 \cdots \alpha_n}{}^* \mid 1 \le \alpha_1 \le M,\ 1 \le \alpha_i \le \chi(\alpha_1,\ldots,\alpha_{i-1}),\ i = 2,\ldots,n,\ \Phi_n(\alpha_1,\ldots,\alpha_n) = x\}$$

and all of these have the same state-distribution $\mathbf{w}_x(n)$. Denote any one of the equivalent subsystems in $C_x(n)$ by $\mathbf{Q}_x(n)$. By definition, let $\mathbf{Q}_1(0) = \mathbf{Q}$. Also let the set of subsystems

$$\{\mathbf{Q}_y(n+1) \mid \mathbf{Q}_y(n+1) \in E(n+1),\ 1 \le y \le E_{n+1},\ y = \Phi_{n+1}(\alpha_1,\ldots,\alpha_n,\beta_{n+1}),\ 1 \le \beta_{n+1} \le \chi(\alpha_1,\ldots,\alpha_n),\ x = \Phi_n(\alpha_1,\ldots,\alpha_n)\}$$

be denoted by $F_x(n)$. In words, the set $F_x(n)$ is the set of subsystems at level (n+1) which are directly related to the subsystem $\mathbf{Q}_x(n)$ at level n, that is, to obtain the distribution $\mathbf{w}_x(n)$, it is required that we have available the set of distributions $\{\mathbf{w}_y(n+1) \mid 1 \le y \le E_{n+1}, \mathbf{Q}_y(n+1) \in F_x(n)\}$. As we have mentioned in the previous section, it may also be possible for the recursion to be cast in terms of the mean performance measures of the subsystems. Another possibility is that it be cast in terms of the normalization constants associated with the state-distributions $\mathbf{w}_x(n)$. In these situations, to obtain the mean performance measures or the normalization constant associated with $\mathbf{Q}_x(n)$, it is required that we have these quantities available for all subsystems in the set $F_x(n)$. These interrelationships are illustrated in Figure 4.5.

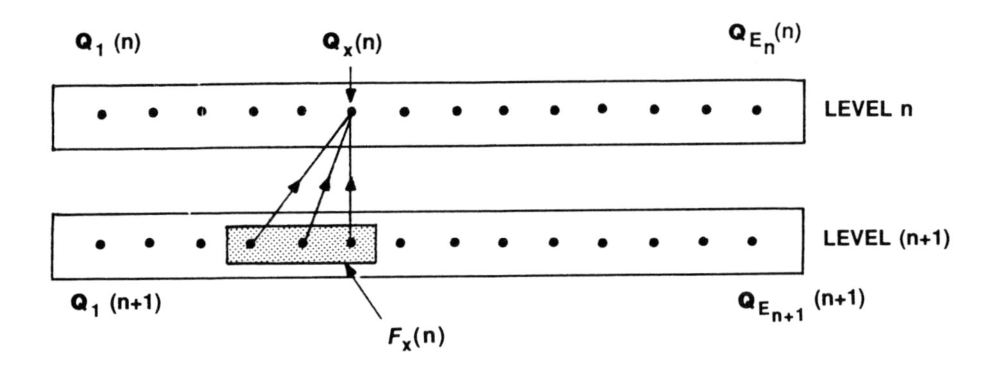

Figure 4.5
Interrelationship Between Subsystems at Adjacent Levels

The interrelationship between the equivalence classes at adjacent levels and the inherent generality of the overall recursion depicted in Fig. 4.4, suggests the following general methodology for the construction of efficient exact computational algorithms. As in Sections 4.1 and 4.2, it is assumed that the BCMP queueing network B(N,**K**) under consideration has been transformed into one that is reversible by performing the transformations which have been enumerated in Section 3.6.

A General Methodology for the Construction of Efficient Exact Computational Algorithms for B(N,K):

Step 1: *L-Level State-Space Partitioning*

(a) Partition the state-space $S^{(R)}$ associated with B(N,**K**) into M sets of communicating states $S(\alpha_1)$, where $1 \le \alpha_1 \le M$.

(b) For levels $n = 1,...,(L-1)$:

For $1 \leq \alpha_1 \leq M$ and $\{\alpha_i \mid 1 \leq \alpha_i \leq \chi(\alpha_1,...,\alpha_{i-1}),\ i = 2,...,n,\ n \geq 2\}$:
Partition $S(\alpha_1,...,\alpha_n)$ into $\chi(\alpha_1,...,\alpha_n)$ sets of communicating states $S(\alpha_1,...,\alpha_n,\alpha_{n+1})$.

Step 2: *Identification of Equivalence Classes*

(a) For levels $n = 1,...,L$:

For $1 \leq \alpha_1 \leq M$ and $\{\alpha_i \mid 1 \leq \alpha_i \leq \chi(\alpha_1,...,\alpha_{i-1}),\ i = 2,...,n,\ n \geq 2\}$:
Determine the state-distribution $\mathbf{v}_{\alpha_1 \cdots \alpha_n}{}^*$ associated with the subsystem $\mathbf{Q}_{\alpha_1 \cdots \alpha_n}{}^*$.

(b) For levels $n = 1,...,L$:

Identify the set $E(n)$ of equivalence classes $C_x(n)$, $1 \leq x \leq E_n$, at level n.

Step 3: *Determination of Recursive Relationships*

(a) For $n = 0,...,(L-1)$:

For $x = 1,...,E_n$:
Find a recursive relationship $R_x(n)$, in terms of mean performance measures or normalization constants, between the subsystem $\mathbf{Q}_x(n)$ and the set of subsystems $F_x(n)$.

In step 3(a), the actual determination of a recursive relationship $R_x(n)$ may be made by appealing to eq. 4.2 which expresses the relationship between the state-distribution of $\mathbf{Q}_x(n)$ and those associated with the set of subsystems $F_x(n)$. Furthermore, since $\mathbf{Q}_x(n)$ and the set of subsystems $F_x(n)$ all define reversible Markov chains, it follows that the reduced system $\mathbf{A}_{\alpha_1 \cdots \alpha_n}$ is, itself, a reversible Markov chain. This is a direct consequence of the condition of detailed balance that exists between all pairs of states contained in $S(\alpha_1,...,\alpha_n)$. Hence, all of the distributions appearing in eq. 4.2 are ones associated with reversible Markov chains and consequently they may all be written as products of terms that may be determined. This information facilitates the determination of a relationship $R_x(n)$.

Having carried out the construction described above, the resulting algorithm for obtaining the mean performance measures is as summarized below:

General Recursive Computational Algorithm for B(N,K)

Initialization:
For $1 \leq x \leq E_L$:
Find the mean performance measures or the normalization constant associated with the subsystem $\mathbf{Q}_x(L)$.

Main Recursion:
For n = (L-1),(L-2),...,1,0:
For $1 \leq x \leq E_n$:
Determine the measures associated with $\mathbf{Q}_x(n)$ using the relationship $R_x(n)$ and the measures associated with the set of systems $F_x(n)$.

The final result in the above general algorithm is the set of measures associated with $\mathbf{Q}_1(0)$ which is, equivalently, the set of measures associated with $\mathbf{Q}$, or $B(N,\mathbf{K})$. If the recursion has been cast in terms of normalization constants, then an additional step is required to determine the mean performance measures of interest in terms of the normalization constants that have been obtained.

The overall computational requirements of the general recursive algorithm described above depend directly on the number of adopted levels L and the number of equivalence classes E_n that may be identified at level n. The space requirement of the algorithm is of the order of

$$\text{Max } \{E_n \mid 1 \le n \le L\},$$

this being the maximum width of the recursive structure, and the time requirement is of the order of

$$\sum_{n=0}^{L} E_n.$$

These requirements depend entirely on the number of equivalence classes to be found in each level of the hierarchy and this number depends on the initial choice of state-space partitioning that is adopted in Steps 1(a) and 1(b) of the constructive methodology.

Although the recursive algorithm that we have formulated is, by nature, very general, a judicious choice of initial state-space partitioning must, of course, be made in order to arrive at an algorithm that can be considered to be computationally efficient. The generality offered by the methodology suggests, however, the existence of a plethora of exact computational algorithms that could, in principle, be developed. Among these algorithms, it is plausible that there exists a number of algorithms that may be considered to be useful from the practical point of view. In the following five sections, we show how the main exact computational algorithms that have been developed may be formulated as special instances of the general methodology that we have proposed. This serves to unify these developments that have hitherto been made. The existence of these algorithms, and the generality offered by our methodology, suggests the existence of other efficient computational algorithms that could possibly be developed. This conjecture is examined in Sections 4.9 and 4.10.

4.4 Constructive Derivation of the Convolution Algorithm

The Convolution Algorithm and the Mean Value Analysis (MVA) algorithm have been the most widely utilized exact algorithms for the solution of product-form queueing networks. The Convolution Algorithm makes use of a recursion in terms of normalization constants, while the MVA algorithm works directly in terms of mean performance measures. The details of these algorithms are the subjects of Sections 5.1 and 5.2, respectively. In this section, we consider the Convolution Algorithm and show how the fundamental underlying recursion may be derived within the context of the general constructive methodology presented in Section 4.3.

Consider a closed multiple-chain BCMP queueing network $B(N,\mathbf{K})$ with state-space $S^{(R)}$ and suppose that in Step 1 of the constructive methodology we choose to adopt the Decomposition by Service Center method, as has been described in Section 3.2. Then, in Step 1(a) of the construction, we partition $S^{(R)}$ into M sets of communicating states $S(\alpha_1)$, $1 \le \alpha_1 \le M$, where $S(\alpha_1) = \{\mathbf{n}^{(R)} \mid \mathbf{n}^{(R)} \in S^{(R)}, \mathbf{n}_N{}^{(R)} = \mathbf{k}^{(1)}\}$, $\mathbf{n}_N{}^{(R)} = (n_{N1},\ldots,n_{NR})$ is the state of node N in $B(N,\mathbf{K})$, $\alpha_1 = \gamma(\mathbf{k}^{(1)})$ and $\gamma(.)$ is a function which maps the elements of the set $\{\mathbf{k}^{(1)} \mid \mathbf{k}^{(1)} = (k_1{}^{(1)},\ldots,k_R{}^{(1)}), \mathbf{0} \le \mathbf{k}^{(1)} \le \mathbf{K}\}$ into the set of integers $\{1,\ldots,M\}$. In this method of state-space partitioning, the partitioning is determined by the state of the Nth node. Consequently, the state-space $S(\alpha_1)$ is exactly the state-space of a queueing network denoted by $B(N\text{-}1,\mathbf{K}\text{-}\mathbf{k}^{(1)})$ and

$$M = \prod_{r=1}^{R} (K_r+1),$$

this being the cardinality of the set $\{\mathbf{k}^{(1)} \mid \mathbf{0} \le \mathbf{k}^{(1)} \le \mathbf{K}\}$.

Now consider level n in Step 1(b) of the construction and suppose that

$$S(\alpha_1,...,\alpha_n) = \{\mathbf{n}^{(R)} \mid \mathbf{n}^{(R)} \in S^{(R)}, \mathbf{n}_N{}^{(R)} = \mathbf{k}^{(1)}, \mathbf{n}_{N-1}{}^{(R)} = \mathbf{k}^{(2)},..., \mathbf{n}_{N-n+1}{}^{(R)} = \mathbf{k}^{(n)}\},$$

where $\mathbf{k}^{(i)} = (k_1{}^{(i)},...,k_R{}^{(i)})$, $\mathbf{0} \le \mathbf{k}^{(1)}+...+\mathbf{k}^{(n)} \le \mathbf{K}$, $\alpha_1 = \gamma(\mathbf{k}^{(1)})$, $\alpha_2 = \gamma(\mathbf{k}^{(2)})$, ..., and $\alpha_n = \gamma(\mathbf{k}^{(n)})$. Following the Decomposition by Service Center method, we partition $S(\alpha_1,...,\alpha_n)$ into $\chi(\alpha_1,...,\alpha_n)$ sets of communicating states $S(\alpha_1,...,\alpha_{n+1})$ according to the state of node (N-n). Hence, $S(\alpha_1,...,\alpha_n)$ is partitioned into the sets of states

$$\{S(\alpha_1,...,\alpha_{n+1}) \mid S(\alpha_1,...,\alpha_n) \supset S(\alpha_1,...,\alpha_{n+1}), \mathbf{n}_{N-n}{}^{(R)} = \mathbf{k}^{(n+1)}, \alpha_{n+1} = \gamma(\mathbf{k}^{(n+1)}), \mathbf{0} \le \mathbf{k}^{(n+1)} \le (\mathbf{K}-\mathbf{k}^{(1)}-...-\mathbf{k}^{(n)})\}.$$

Consequently,

$$\chi(\alpha_1,...,\alpha_n) = \prod_{r=1}^{R} (K_r-k_r{}^{(1)}-...-k_r{}^{(n)}+1)$$

and the set of states $S(\alpha_1,...,\alpha_{n+1})$ corresponds exactly with the state-space of a queueing network denoted by $B(N\text{-}n\text{-}1,\mathbf{K}-\mathbf{k}^{(1)}-...-\mathbf{k}^{(n+1)})$. Continuing this method of state-space partitioning, the total number of levels L that result is L = N, where N is the total number of nodes in $B(N,\mathbf{K})$.

Having partitioned $S^{(R)}$ into L levels of subsets $S(\alpha_1,...,\alpha_n)$ and recognized that $S(\alpha_1,...,\alpha_n)$ is the state-space associated with the queueing network $B(N\text{-}n,\mathbf{K}-\mathbf{k}^{(1)}-...-\mathbf{k}^{(n)})$, where $\alpha_i = \gamma(\mathbf{k}^{(i)})$, the second step in the construction is to find the state-distribution $\mathbf{v}_{\alpha_1\cdots\alpha_n}{}^*$ associated with $\mathbf{Q}_{\alpha_1\cdots\alpha_n}{}^*$ and identify the equivalence classes $C_x(n)$. The distribution $\mathbf{v}_{\alpha_1\cdots\alpha_n}{}^*$ is known since the subsystem $\mathbf{Q}_{\alpha_1\cdots\alpha_n}{}^*$ may be identified as a queueing network denoted by $B(N\text{-}n,\mathbf{K}-\mathbf{k}^{(1)}-...-\mathbf{k}^{(n)})$. The state of this system is $(\mathbf{n}_1{}^{(R)},...,\mathbf{n}_{N-n}{}^{(R)})$ and the state-distribution is, according to eq. 2.10,

$$\pi(\mathbf{n}_1{}^{(R)},...,\mathbf{n}_{N-n}{}^{(R)}) = C^{-1}G_1{}^{-1} \{\prod_{i=1}^{N-n} f_i(\mathbf{n}_i{}^{(R)})\}\{\prod_{i=N-n+1}^{N} f_i(\mathbf{k}^{(N-i+1)})\},$$

where C is a normalization constant. The above may be written in the form

$$\pi(\mathbf{n}_1^{(R)},\ldots,\mathbf{n}_{N-n}^{(R)}) = G_{N-n}(\mathbf{K},\mathbf{k}^{(1)},\ldots,\mathbf{k}^{(n)})^{-1} \prod_{i=1}^{N-n} f_i(\mathbf{n}_i^{(R)}), \tag{4.3}$$

where, by definition,

$$G_{N-n}(\mathbf{K},\mathbf{k}^{(1)},\ldots,\mathbf{k}^{(n)}) = \sum_{\mathbf{n}^{(R)} \in S(\alpha_1,\ldots,\alpha_n)} \prod_{i=1}^{N-n} f_i(\mathbf{n}_i^{(R)}).$$

We now identify the equivalence classes to be found at level n when the adopted state-space partitioning follows that of the Decomposition by Service Center method. Two subsystems $\mathbf{Q}_{\alpha_1\cdots\alpha_n}{}^*$ and $\mathbf{Q}_{\beta_1\cdots\beta_n}{}^*$, where $\alpha_1 = \gamma(\mathbf{a}^{(1)})$, ..., $\alpha_n = \gamma(\mathbf{a}^{(n)})$, $\beta_1 = \gamma(\mathbf{b}^{(1)})$, ..., $\beta_n = \gamma(\mathbf{b}^{(n)})$, $\mathbf{a}^{(i)} = (a_1{}^{(i)},\ldots,a_R{}^{(i)})$ and $\mathbf{b}^{(i)} = (b_1{}^{(i)},\ldots,b_R{}^{(i)})$, belong to the same equivalence class $C_x(n)$ if $\mathbf{v}_{\alpha_1\cdots\alpha_n}{}^* = \mathbf{v}_{\beta_1\cdots\beta_n}{}^*$. This condition is ensured under the simple condition that

$$(\mathbf{a}^{(1)} + \ldots + \mathbf{a}^{(n)}) = (\mathbf{b}^{(1)} + \ldots + \mathbf{b}^{(n)}).$$

In other words, the two subsystems $\mathbf{Q}_{\alpha_1\cdots\alpha_n}{}^*$ and $\mathbf{Q}_{\beta_1\cdots\beta_n}{}^*$ are equivalent, as far as their equilibrium state-distributions are concerned, if the population vector of $B(N-n,\mathbf{K}-\mathbf{a}^{(1)}-\ldots-\mathbf{a}^{(n)})$ is the same as that of $B(N-n,\mathbf{K}-\mathbf{b}^{(1)}-\ldots-\mathbf{b}^{(n)})$. Since at the *n*th level in the method of hierarchical state-space partitioning that has been adopted $\mathbf{0} \le \mathbf{k}^{(i)} \le \mathbf{K}$, for $1 \le i \le n$, and $\mathbf{0} \le (\mathbf{k}^{(1)}+\ldots+\mathbf{k}^{(n)}) \le \mathbf{K}$, the number of distinct equivalence classes E_n at level n is

$$E_n = \prod_{r=1}^{R} (K_r+1),$$

this being the cardinality of the set

$$\{(\mathbf{k}^{(1)}+\ldots+\mathbf{k}^{(n)}) \mid \mathbf{0} \le \mathbf{k}^{(i)} \le \mathbf{K}, \text{ for } 1 \le i \le n,\ \mathbf{0} \le \mathbf{k}^{(1)}+\ldots+\mathbf{k}^{(n)} \le \mathbf{K}\}.$$

At the first level, we have $E_1 = \Omega(1)$. However, for $n \geq 2$, the number of equivalence classes at level n is very much smaller that the width $\Omega(n)$. The width is given explicitly by eq. 4.1 with

$$M = \prod_{r=1}^{R} (K_r+1),$$

$$\chi(\alpha_1) = \prod_{r=1}^{R} (K_r-k_r^{(1)}+1), \qquad \chi(\alpha_1,\alpha_2) = \prod_{r=1}^{R} (K_r-k_r^{(1)}-k_r^{(2)}+1),$$

and so on.

The third step in the constructive methodology is to establish a recursive relationship $R_x(n)$ between the subsystem $\mathbf{Q}_x(n)$, corresponding to the equivalence class $C_x(n)$, and the set of subsystems $F_x(n)$. In the Convolution Algorithm, the recursion which is utilized is one in terms of normalization constants. To establish this recursion, let us first find a relationship in terms of the state-distribution $\mathbf{w}_x(n)$, corresponding to the equivalence class $C_x(n)$, and the distributions associated with $F_x(n)$. The system $\mathbf{Q}_x(n)$ is a queueing network denoted by $B(N\text{-}n,\mathbf{K}\text{-}\mathbf{k}^{(1)}\text{-}\ldots\text{-}\mathbf{k}^{(n)})$, such that $x = \Phi_n(\alpha_1,\ldots,\alpha_n)$, where $\alpha_i = \gamma(\mathbf{k}^{(i)})$. The state-distribution for this system is given by eq. 4.3. Now consider the state-distributions of the systems in the set $F_x(n)$, where

$$F_x(n) = \{B(N\text{-}n\text{-}1,\mathbf{K}\text{-}\mathbf{k}^{(1)}\text{-}\ldots\text{-}\mathbf{k}^{(n)}\text{-}\mathbf{p}) \mid \mathbf{p} = (p_1,\ldots,p_R),\ \mathbf{0} \leq \mathbf{p} \leq \mathbf{K}\text{-}\mathbf{k}^{(1)}\text{-}\ldots\text{-}\mathbf{k}^{(n)}\}.$$

The state-distribution $\pi_{\mathbf{p}}(\mathbf{n}_1^{(R)},\ldots,\mathbf{n}_{N-n-1}^{(R)})$ associated with $B(N\text{-}n\text{-}1, \mathbf{K}\text{-}\mathbf{k}^{(1)}\text{-}\ldots\text{-}\mathbf{k}^{(n)}\text{-}\mathbf{p})$ is

$$\pi_{\mathbf{p}}(\mathbf{n}_1^{(R)},\ldots,\mathbf{n}_{N-n-1}^{(R)}) = G_{N-n-1}(\mathbf{K},\mathbf{k}^{(1)},\ldots,\mathbf{k}^{(n)},\mathbf{p})^{-1} \prod_{i=1}^{N-n-1} f_i(\mathbf{n}_i^{(R)}), \qquad (4.4)$$

where, by definition,

$$G_{N-n-1}(\mathbf{K},\mathbf{k}^{(1)},\ldots,\mathbf{k}^{(n)},\mathbf{p}) = \sum_{\substack{\mathbf{n}^{(R)} \in S(\alpha_1,\ldots,\alpha_n) \\ \mathbf{n}_{N-n}^{(R)} = \mathbf{p}}} \prod_{i=1}^{N-n-1} f_i(\mathbf{n}_i^{(R)}).$$

We may now write the relationship between $\pi(\mathbf{n}_1^{(R)},\ldots,\mathbf{n}_{N-n}^{(R)})$, for the system $B(N\text{-}n,\mathbf{K}\text{-}\mathbf{k}^{(1)}\text{-}\ldots\text{-}\mathbf{k}^{(n)})$, and $\pi_{\mathbf{p}}(\mathbf{n}_1^{(R)},\ldots,\mathbf{n}_{N-n-1}^{(R)})$. By definition, we have

$$\pi(\mathbf{n}_1^{(R)},\ldots,\mathbf{n}_{N-n-1}^{(R)},\mathbf{p}) = \pi_{\mathbf{p}}(\mathbf{n}_1^{(R)},\ldots,\mathbf{n}_{N-n-1}^{(R)}) \ldots$$

$$\sum_{\substack{(\mathbf{n}_1^{(R)},\ldots,\mathbf{n}_{N-n}^{(R)}) \in S(\alpha_1,\ldots,\alpha_n) \\ \mathbf{n}_{N-n}^{(R)} = \mathbf{p}}} \pi(\mathbf{n}_1^{(R)},\ldots,\mathbf{n}_{N-n-1}^{(R)},\mathbf{n}_{N-n}^{(R)}),$$

where $\alpha_1 = \gamma(\mathbf{k}^{(1)})$, ..., and $\alpha_n = \gamma(\mathbf{k}^{(n)})$. Hence, using eqs. 4.3 and 4.4, we have

$$f_{N-n}(\mathbf{p})G_{N-n-1}(\mathbf{K},\mathbf{k}^{(1)},\ldots,\mathbf{k}^{(n)},\mathbf{p}) \,/\, G_{N-n}(\mathbf{K},\mathbf{k}^{(1)},\ldots,\mathbf{k}^{(n)}) =$$

$$\sum_{\substack{(\mathbf{n}_1^{(R)},\ldots,\mathbf{n}_{N-n}^{(R)}) \in S(\alpha_1,\ldots,\alpha_n) \\ \mathbf{n}_{N-n}^{(R)} = \mathbf{p}}} \pi(\mathbf{n}_1^{(R)},\ldots,\mathbf{n}_{N-n}^{(R)}),$$

but

$$\sum_{\mathbf{0} \le \mathbf{p} \le (\mathbf{K}\text{-}\mathbf{k}^{(1)}\text{-}\ldots\text{-}\mathbf{k}^{(n)})} \ldots \sum_{\substack{(\mathbf{n}_1^{(R)},\ldots,\mathbf{n}_{N-n}^{(R)}) \in S(\alpha_1,\ldots,\alpha_n) \\ \mathbf{n}_{N-n}^{(R)} = \mathbf{p}}} \pi(\mathbf{n}_1^{(R)},\ldots,\mathbf{n}_{N-n}^{(R)}) = 1,$$

so that we find

$$G_{N-n}(\mathbf{K},\mathbf{k}^{(1)},\ldots,\mathbf{k}^{(n)}) =$$

$$\sum_{\mathbf{0} \le \mathbf{p} \le (\mathbf{K}\text{-}\mathbf{k}^{(1)}\text{-}\ldots\text{-}\mathbf{k}^{(n)})} f_{N-n}(\mathbf{p})G_{N-n-1}(\mathbf{K},\mathbf{k}^{(1)},\ldots,\mathbf{k}^{(n)},\mathbf{p}). \qquad (4.5)$$

Noting that $G_{N-n}(\mathbf{K},\mathbf{k}^{(1)},...,\mathbf{k}^{(n)})$ and $G_{N-n-1}(\mathbf{K},\mathbf{k}^{(1)},...,\mathbf{k}^{(n)},\mathbf{p})$ depend only on their arguments through $(\mathbf{K}-\mathbf{k}^{(1)}-...-\mathbf{k}^{(n)})$ and $(\mathbf{K}-\mathbf{k}^{(1)}-...-\mathbf{k}^{(n)}-\mathbf{p})$, respectively, we may write eq. 4.5 in the more concise form

$$G_m(\mathbf{x}) = \sum_{\mathbf{0} \le \mathbf{p} \le \mathbf{x}} f_m(\mathbf{p}) G_{m-1}(\mathbf{x}-\mathbf{p}), \tag{4.6}$$

where $m = (N-n)$ and $\mathbf{x} = (\mathbf{K}-\mathbf{k}^{(1)}-...-\mathbf{k}^{(n)})$.

Equation 4.6, which expresses a multi-dimensional convolution operation, is the recursive equation that forms the fundamental basis of the Convolution Algorithm. It provides a relationship between the normalization constant of the subsystem $\mathbf{Q}_x(n)$ at level n in the hierarchical decomposition and the normalization constants associated with the subsystems at level (n+1) in the set $F_x(n)$. In the Convolution Algorithm, which is to be considered in more detail in the following chapter, the quantity $G_N(\mathbf{K})$ is computed using eq. 4.6 with the initial conditions $G_1(\mathbf{x}) = 1$, for $\mathbf{0} \le \mathbf{x} \le \mathbf{K}$. Having computed $G_N(\mathbf{K})$, it is then possible to obtain immediately the mean performance measures of interest. In view of the above constructive steps that we have taken to arrive at eq. 4.6, we see that the Convolution Algorithm is merely a special instance of the general constructive methodology that we have proposed. It is the algorithm which results when we define the recursive scheme in terms of normalization constants and when we choose to decompose the state-space $S^{(R)}$ according to the Decomposition by Service Center method. Since the number of equivalence classes in each level of the hierarchy is

$$E_n = \prod_{r=1}^{R} (K_r+1)$$

and the number of levels L is N, it follows that the space and time requirements of the Convolution Algorithm are of the order of

$$\prod_{r=1}^{R} (K_r+1) \quad \text{and} \quad N\prod_{r=1}^{R} (K_r+1),$$

respectively. These requirements are, of course, a direct consequence of adopting the Decomposition by Service Center method for the state-space partitioning.

4.5 Constructive Derivation of MVA

We now consider the derivation of the MVA algorithm within the context of the methodology developed in Section 4.3. We first consider the case of constant-speed service-rate functions. In this case, the fundamental recursive difference equation, which is utilized in MVA, is a relation between the mean waiting-times in the system $B(N,\mathbf{K})$ and the mean queue-lengths in the set of systems $\{B(N,\mathbf{K}-\mathbf{1}_r) \mid 1 \leq r \leq R\}$. In the literature, the recursive equation which forms the basis of the MVA algorithm is usually derived from the *Arrival Theorem* [KEL3,LAV1,SEV1] for product-form queueing networks, as stated informally below.

The Arrival Theorem: The state-distribution for $B(N,\mathbf{K})$ at arrival epochs of type r customers is equal to the state-distribution of $B(N,\mathbf{K}-\mathbf{1}_r)$ at arbitrary times.

Here we shall give a derivation of the recursion employed in MVA within the context of the methodology that we have described. Our development will be based on a set of interrelated 'parallel' decompositions.

Consider the queueing network $B(N,\mathbf{K})$ and suppose, initially, that we have constant-speed service-rate functions at the nodes of the network. Also suppose that there are no IS queues in the network. The case in which there are IS queues will be treated when we consider the situation of state-dependent service-rate functions. Now consider partitioning the state-space according to the disposition of one particular customer of chain r. Without loss of generality, consider the K_Rth customer of chain R and denote its state in the network by the vector random variable **c**, where $\mathbf{c} = (c_1,...,c_N)$ and

$$c_i = \begin{cases} 1, \text{ if the } K_R\text{th customer of chain R is at node i}, \\ 0, \text{ otherwise.} \end{cases}$$

Since we *distinguish* the K_Rth customer of chain R, the state-space of the network is

$$S^{(R)'} = \{(\mathbf{n}^{(R)},\mathbf{c}) \mid \mathbf{c} = (c_1,...,c_N),\ 0 \le n_{ir} \le K_r \text{ for } 1 \le r \le R\text{-}1,$$
$$0 \le n_{iR} \le K_R\text{-}1,\ \sum_{i=1}^{N} n_{ir} = K_r \text{ for } r = 1,...,R\text{-}1,\ \sum_{i=1}^{N} n_{iR} = K_R\text{-}1,\ \mathbf{c} = \mathbf{1}_i$$
$$\text{for } 1 \le i \le N\}.$$

Let $S(\alpha_1) = \{(\mathbf{n}^{(R)},\mathbf{c}) \mid (\mathbf{n}^{(R)},\mathbf{c}) \in S^{(R)'},\ c_i = 0 \text{ for } i \neq \alpha_1,\ c_{\alpha_1} = 1\}$, where $1 \le \alpha_1 \le M$ and $M = N$. In words, $S(\alpha_1)$ is the set of states in which the K_Rth customer of chain r is at node α_1. In this situation, the subsystem $\mathbf{Q}_{\alpha_1}{}^*$ may be characterized as a queueing network of type $B(N,\mathbf{K}\text{-}\mathbf{1}_R)$ in which there is a single so-called 'single-customer self-looping' (SCSL) chain at node α_1. The reduced system **A** may be characterized as a network that contains only the K_Rth customer of chain R.

Now let us proceed to establish a relationship $R_1(0)$ between the system $\mathbf{Q}_1(0)$, corresponding to the queueing network $B(N,\mathbf{K})$, and the subsystems $\mathbf{Q}_{\alpha_1}(1)$ corresponding to $\mathbf{Q}_{\alpha_1}{}^*$ at the first level. Consider the state-distribution **u** of the reduced system **A**, where $\mathbf{u} = (u_1,...,u_N)$. The element u_i is the probability that the K_Rth customer of chain R is at node i in $B(N,\mathbf{K})$. By definition, we have the relationship

$$\pi(\mathbf{n}^{(R)} \text{ in } \mathbf{Q}_1(0),\mathbf{c} = \mathbf{1}_i) = \pi(\mathbf{n}^{(R)} \text{ in } \mathbf{Q}_i(1))u_i. \tag{4.7}$$

It follows from eq. 2.10 that, in the above,

$$\pi(\mathbf{n}^{(R)} \text{ in } \mathbf{Q}_1(0), \mathbf{c} = \mathbf{1}_i) = G_N(\mathbf{K})'^{-1}(n_i+1) w_{iR} \prod_{j=1}^{N} n_j! \left(\prod_{r=1}^{R} w_{jr}{}^{n_{jr}}/n_{jr}! \right), \tag{4.8}$$

where

$$G_N(\mathbf{K})' = \sum_{\substack{(\mathbf{n}^{(R)},\mathbf{c}) \in S^{(R)'} \\ \mathbf{c} = \mathbf{1}_i \\ 1 \le i \le N}} (n_i+1) w_{iR} \left\{ \prod_{j=1}^{N} n_j! \left(\prod_{r=1}^{R} w_{jr}{}^{n_{jr}}/n_{jr}! \right) \right\}.$$

We also have that

$$\pi(\mathbf{n}^{(R)} \text{ in } \mathbf{Q}_i(1)) = G_N(\mathbf{K},i)^{-1}(n_i+1) \prod_{j=1}^{N} n_j! \left(\prod_{r=1}^{R} w_{jr}{}^{n_{jr}}/n_{jr}! \right), \tag{4.9}$$

where

$$G_N(\mathbf{K},i) = \sum_{\substack{(\mathbf{n}^{(R)},\mathbf{c}) \in S^{(R)'} \\ \mathbf{c} = \mathbf{1}_i}} (n_i+1) \prod_{j=1}^{N} n_j! \left(\prod_{r=1}^{R} w_{jr}{}^{n_{jr}}/n_{jr}! \right). \tag{4.10}$$

Substituting eqs. 4.8 and 4.9 into eq. 4.7, we obtain

$$u_i = w_{iR} G_N(\mathbf{K},i) \,/\, G_N(\mathbf{K})'. \tag{4.11}$$

Equation 4.11 expresses a relation among the marginal probability u_i that the K_Rth customer of chain R in B(N,**K**) is at node i and the normalization constants associated with the systems $\mathbf{Q}_1(0)$ and $\mathbf{Q}_i(1)$. Now since $\sum_{i=1}^{N} u_i = 1$, we have that

$$G_N(\mathbf{K})' = \sum_{i=1}^{N} w_{iR} G_N(\mathbf{K},i) \tag{4.12}$$

and from eq. 4.10, we see that

$$G_N(\mathbf{K},i) = G_N(\mathbf{K}-\mathbf{1}_R)(Q_i(\mathbf{K}-\mathbf{1}_R) + 1), \tag{4.13}$$

where $G_N(\mathbf{K}-\mathbf{1}_R)$ is the normalization constant associated with $B(N,\mathbf{K}-\mathbf{1}_R)$ and $Q_i(\mathbf{K}-\mathbf{1}_R)$ is the mean queue-length at node i in $B(N,\mathbf{K}-\mathbf{1}_R)$. Hence, substituting eq. 4.13 into eq. 4.12, we obtain

$$G_N(\mathbf{K})' = \sum_{i=1}^{N} w_{iR} G_N(\mathbf{K}-\mathbf{1}_R)(Q_i(\mathbf{K}-\mathbf{1}_R) + 1).$$

But it is well-known (see e.g. [BRU1]) that, in terms of the notation adopted here,

$$T_{iR}(\mathbf{K}) = e_{iR} K_R G_N(\mathbf{K}-\mathbf{1}_R) \,/\, G_N(\mathbf{K})'$$

so that

$$T_{iR}(\mathbf{K}) = e_{iR} K_R \,/ \sum_{j=1}^{N} w_{jR}(Q_j(\mathbf{K}-\mathbf{1}_R) + 1). \tag{4.14}$$

We also have, by definition, that $Q_{iR}(\mathbf{K}) = K_R u_i$ so that, using eqs. 4.11 and 4.13,

$$Q_{iR}(\mathbf{K}) = K_R w_{iR} G_N(\mathbf{K}-\mathbf{1}_R)(Q_i(\mathbf{K}-\mathbf{1}_R) + 1) \,/\, G_N(\mathbf{K})'.$$

Hence,

$$Q_{iR}(\mathbf{K}) = t_{iR} T_{iR}(\mathbf{K})(Q_i(\mathbf{K}-\mathbf{1}_R) + 1). \tag{4.15}$$

Using Little's result [LIT1], eq. 4.15 may be rewritten as

$$W_{iR}(\mathbf{K}) = t_{iR}(Q_i(\mathbf{K}-\mathbf{1}_R) + 1), \tag{4.16}$$

where $W_{iR}(\mathbf{K})$ is the mean waiting-time for customers of chain R at node i in $B(N,\mathbf{K})$. Eq. 4.16 is the fundamental waiting-time equation that forms the basis of the MVA algorithm in the case of constant service-rate functions. Eqs. 4.14 and 4.15 may be viewed to be a direct consequence of eq. 4.16 and Little's result [LIT1].

Equations 4.14, 4.15 and 4.16, which constitute the fundamental recursive equations of MVA in the case of constant-speed service-rate functions, relate the mean performance measures for chain R in the network $B(N,\mathbf{K})$ to those of the network $B(N,\mathbf{K}-\mathbf{1}_R)$. These relationships were obtained by considering a single-level state-space partitioning conditioned on the position of the K_rth customer of chain R in the network. To obtain relationships for the other routing chains, we can consider the application of R *parallel* single-level decompositions, where in the rth decomposition, $1 \leq r \leq R$, we condition the state-space partitioning of $S^{(R)}$ on the position of the K_rth customer of chain r. Hence, in the rth decomposition we obtain the relationships between the measures associated with $B(N,\mathbf{K})$ and those of $B(N,\mathbf{K}-\mathbf{1}_r)$. These relationships are the same as eqs. 4.14, 4.15 and 4.16, but with R replaced by r. At the first level in the parallel decomposition, the set of systems is, therefore, $\{B(N,\mathbf{K}-\mathbf{1}_r) \mid 1 \leq r \leq R\}$.

If we now apply the same type of parallel single-level decomposition to each of the systems obtained in the first level, the set of systems which result at the second level is $\{B(N,\mathbf{K}-\mathbf{1}_{a_1}-\mathbf{1}_{a_2}) \mid 1 \leq a_1,a_2 \leq R\}$. In general, at the nth level, the set of systems obtained is $\{B(N,\mathbf{K}-\mathbf{1}_{a_1}-\mathbf{1}_{a_2}-\ldots-\mathbf{1}_{a_n}) \mid (\mathbf{K}-\mathbf{1}_{a_1}-\ldots-\mathbf{1}_{a_n}) \geq \mathbf{0},\ 1 \leq a_1,a_2,\ldots,a_n \leq R\}$. This multiple-level interrelationship of systems is illustrated in Figure 4.6. It is this hierarchy, together with the set of relationships provided by eqs. 4.14 to 4.16, that constitute the MVA algorithm in the case of constant-speed service-rate functions.

We now examine the equivalence classes that exist at level n in the hierarchy. Two systems $B(N,\mathbf{K}-\mathbf{1}_{a_1}-...-\mathbf{1}_{a_n})$ and $B(N,\mathbf{K}-\mathbf{1}_{b_1}-...-\mathbf{1}_{b_n})$ at level n belong to the same equivalence class if $(\mathbf{K}-\mathbf{1}_{a_1}-...-\mathbf{1}_{a_n}) = (\mathbf{K}-\mathbf{1}_{b_1}-...-\mathbf{1}_{b_n})$. Hence, the set of distinguishable systems at level n is $\{B(N,\mathbf{k}) \mid 0 \le k_r \le K_r, \sum_{r=1}^{R} k_r = K_1+...+K_R-n\}$. Consequently, the width E_n of level n in the hierarchy is the number of compositions of n into exactly R parts c_r, $1 \le r \le R$, subject to the constraint that $0 \le c_r \le K_r$. Denote this number by $C_{\mathbf{K}}(n)$. It now follows that the space requirement of the MVA algorithm is of the order of

$$\text{Max}\ \{NRC_{\mathbf{K}}(n) \mid 1 \le n \le (K_1+...+K_R)\}.$$

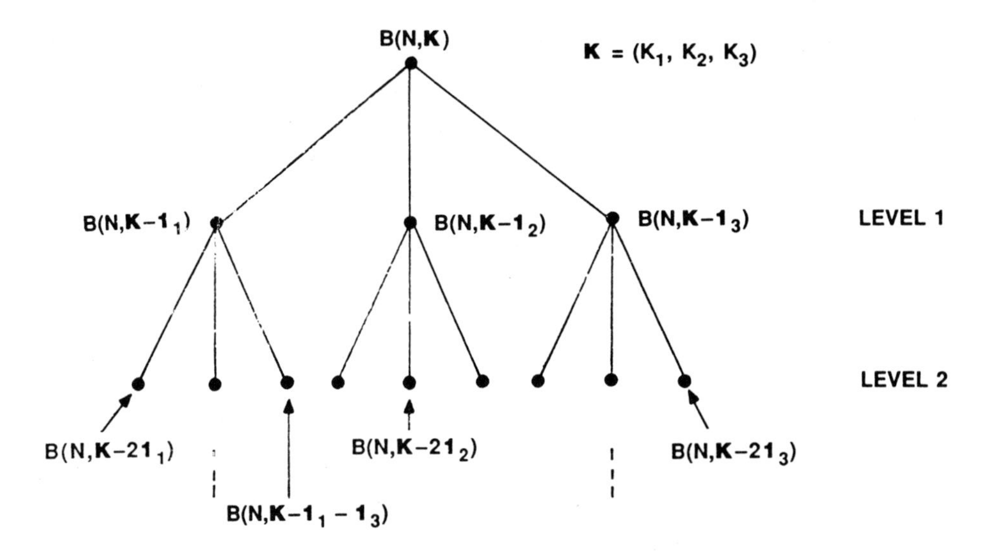

Figure 4.6
Interrelationships Between Systems in the MVA Recursion

Since the overall recursion of the hierarchical structure is over the set of population vectors $\{\mathbf{k} \mid \mathbf{0} \le \mathbf{k} \le \mathbf{K}\}$, the time requirement is of the order of

$$NR \prod_{r=1}^{R} (K_r+1).$$

In the literature, the space requirement is usually given to be of the order of

$$NR \prod_{r=1}^{R-1} (K_r+1).$$

This is the case when the recursion over $\mathbf{k}$ is performed in the following straightforward manner:

For $k_R = 0,\ldots,K_R$:
For $k_{R-1} = 0,\ldots,K_{R-1}$:
•
•
•
For $k_2 = 0,\ldots,K_2$:
For $k_1 = 0,\ldots,k_1$:
$\mathbf{k} = (k_1,k_2,\ldots,k_{R-1},k_R)$

We now consider the development of the fundamental underlying recursive equation of MVA in the case of state-dependent service-rate functions of the form $\mu_i(n_i)$. In this case, the recursive structure is the same as that developed above for the constant service-rate case but the recursive relationship is cast in terms of marginal queue-length distributions.

Let $P_i(x,\mathbf{K})$ denote the probability that there are a total of x customers at node i in $B(N,\mathbf{K})$. By definition,

$P_i(x,\mathbf{K},\mathbf{c} = \mathbf{1}_i) = P_i(x,\mathbf{K} \mid \mathbf{c} = \mathbf{1}_i)\, u_i,$

where u_i is the probability that the K_Rth customer of chain R is at node i and $\mathbf{c}$ is as defined above. From eq. 2.10, it follows that

$$P_i(x,\mathbf{K} \mid \mathbf{c} = \mathbf{1}_i) = \sum_{\substack{(\mathbf{n}^{(R)},\mathbf{c}) \in S^{(R)'} \\ n_i = x-1 \\ \mathbf{c} = \mathbf{1}_i}} G_N(\mathbf{K})'^{-1}((n_i+1)w_{iR}/\mu_i(n_i+1)) \{\prod_{j=1}^{N} n_j!(1/\prod_{a=1}^{n_j}\mu_j(a))\prod_{r=1}^{R} w_{jr}^{n_{jr}}/n_{jr}!\}C(i)^{-1},$$

where $S^{(R)'}$ is as defined above,

$$C(i) = \sum_{\substack{(\mathbf{n}^{(R)},\mathbf{c}) \in S^{(R)'} \\ \mathbf{c} = \mathbf{1}_i}} G_N(\mathbf{K})'^{-1}((n_i+1)w_{iR}/\mu_i(n_i+1)) \{\prod_{j=1}^{N} n_j!(1/\prod_{a=1}^{n_j}\mu_j(a))\prod_{r=1}^{R} w_{jr}^{n_{jr}}/n_{jr}!\}$$

and

$$G_N(\mathbf{K})' = \sum_{(\mathbf{n}^{(R)},\mathbf{c}) \in S^{(R)'}} ((n_i+1)w_{iR}/\mu_i(n_i+1))\{\prod_{j=1}^{N} n_j!(1/\prod_{a=1}^{n_j}\mu_j(a))\prod_{r=1}^{R} w_{jr}^{n_{jr}}/n_{jr}!\}.$$

Hence,

$$P_i(x,\mathbf{K} \mid \mathbf{c} = \mathbf{1}_i) = G_N(\mathbf{K})'^{-1}(xw_{iR}/\mu_i(x))P_i(x-1,\mathbf{K}-\mathbf{1}_R)G_N(\mathbf{K}-\mathbf{1}_R)C(i)^{-1}.$$

By definition, $u_i = C(i)$ so that

$$\begin{aligned} P_i(x,\mathbf{K},\mathbf{c} = \mathbf{1}_i) &= G_N(\mathbf{K})'^{-1}(xw_{iR}/\mu_i(x))P_i(x-1,\mathbf{K}-\mathbf{1}_R)G_N(\mathbf{K}-\mathbf{1}_R) \\ &= T_{iR}(\mathbf{K})(xt_{iR}/\mu_i(x))P_i(x-1,\mathbf{K}-\mathbf{1}_R)/K_R. \end{aligned}$$

Summing both sides of the above over x, $0 \le x \le K$, we obtain

$$\pi(\mathbf{c} = \mathbf{1}_i) = (t_{iR}T_{iR}(\mathbf{K})/K_R) \sum_{k=1}^{K} kP_i(k-1,\mathbf{K}-\mathbf{1}_R)/\mu_i(k),$$

where $K = K_1+\ldots+K_R$. But $Q_{iR}(\mathbf{K}) = K_R\pi(\mathbf{c}=\mathbf{1}_i)$ and $W_{iR}(\mathbf{K}) = Q_{iR}(\mathbf{K})/T_{iR}(\mathbf{K})$ so that

$$W_{iR}(\mathbf{K}) = t_{iR} \sum_{k=1}^{K} kP_i(k-1,\mathbf{K}-\mathbf{1}_R)/\mu_i(k). \tag{4.17}$$

Equation 4.17 is the fundamental waiting-time equation of the MVA algorithm in the case of state-dependent service-rate functions. In the case that node i is an IS queue, we have $\mu_i(k) = k$ and eq. 4.17 reduces to $W_{iR}(\mathbf{K}) = t_{iR}$, as expected. To obtain the corresponding waiting-time relationships for other routing chains in $B(N,\mathbf{K})$, it suffices to replace R by r in eq. 4.17.

4.6 Constructive Derivation of RECAL

In this Section, we consider the constructive derivation of the Recursion by Chain Algorithm (RECAL). The distinguishing feature of this algorithm relative to the Convolution Algorithm and MVA is that the computational and storage requirements are polynomial in the number of closed routing chains R. Consequently, RECAL is computationally attractive for the solution of queueing networks in which the number of closed routing chains is large relative to the number of nodes N in the network. In [CON3], it has been established that RECAL offers a significant advantage when R is greater than approximately twice the number of nodes N. The recursion employed in RECAL is intimately related to the method of Decomposition by Routing Chain. Indeed, it is the algorithm which results from the constructive methodology of Section 4.3 when the adopted state-space partitioning is the hierarchical Decomposition by Routing Chain procedure described previously in Section 3.4.

Consider the queueing network $B(N,\mathbf{K})$ with state-space $S^{(R)}$ and suppose, initially, that we have constant-speed service-rate functions at the nodes of the network. Furthermore, suppose that in Step 1(a) of the constructive methodology we partition $S^{(R)}$ into M sets of communicating states $S(\alpha_1)$, $1 \leq \alpha_1 \leq M$, where

$$S(\alpha_1) = \{\mathbf{n}^{(R)} \mid \mathbf{n}^{(R)} \in S^{(R)}, \mathbf{n}_R = \mathbf{k}_R, \alpha_1 = \zeta^{(R)}(\mathbf{k}_R)\},$$

$\mathbf{n}_R = (n_{1R},\ldots,n_{NR})$, $\mathbf{k}_R = (k_{1R},\ldots,k_{NR})$ and $\zeta^{(r)}(.)$, $1 \leq r \leq R$, is a function which maps the set of vectors

$$\{\mathbf{k}_r \mid k_{ir} \geq 0, 1 \leq i \leq N, \sum_{i=1}^{N} k_{ir} = K_r\}$$

into the set of integers

$$\{1,\ldots,\binom{K_r+N-1}{N-1}\}.$$

The vector $\mathbf{n}_R$ is the state of the customers of chain R in the network $B(N,\mathbf{K})$. It may also be seen that the set of states $S(\alpha_1)$ corresponds exactly with the state-space of a queueing network denoted by $M(R-1,\mathbf{k}_R)$, as has been defined in Section 3.4. We also have that

$$M = \binom{K_R+N-1}{N-1}.$$

Now consider Step 1(b) of the construction and suppose that

$$S(\alpha_1,\ldots,\alpha_n) = \{\mathbf{n}^{(R)} \mid \mathbf{n}^{(R)} \in S^{(R)}, \mathbf{n}_R = \mathbf{k}_R, \mathbf{n}_{R-1} = \mathbf{k}_{R-1},\ldots,\mathbf{n}_{R-n+1} = \mathbf{k}_{R-n+1}\},$$

where $\alpha_r = \zeta^{(r)}(\mathbf{k}_r)$ for $(R-n+1) \leq r \leq R$. Following the Decomposition by Routing Chain method, we partition $S(\alpha_1,\ldots,\alpha_n)$ into $\chi(\alpha_1,\ldots,\alpha_n)$ sets of communicating states $S(\alpha_1,...,\alpha_{n+1})$ according to the state of the customers of routing chain (R-n). Hence, for $n \geq 1$,

$$S(\alpha_1,...,\alpha_n,\alpha_{n+1}) = \{\mathbf{n}^{(R)} \mid \mathbf{n}^{(R)} \in S(\alpha_1,...,\alpha_n),\ \mathbf{n}_{R-n} = \mathbf{k}_{R-n}\},$$

where $\zeta^{(R-n)}(\mathbf{k}_{R-n}) = \alpha_{n+1}$. We also have that

$$\chi(\alpha_1,...,\alpha_n) = \binom{K_{R-n}+N-1}{N-1}.$$

The set of states $S(\alpha_1,...,\alpha_{n+1})$ corresponds exactly with the state-space of the queueing network $M(R\text{-}n\text{-}1,\mathbf{k}_R+...+\mathbf{k}_{R-n})$. The total number of levels L involved in the hierarchical decomposition is L = R.

Having partitioned $S^{(R)}$ into L levels of subsets $S(\alpha_1,...,\alpha_n)$, the second step in the construction is to find the state-distribution $\mathbf{v}_{\alpha_1...\alpha_n}{}^*$ associated with $\mathbf{Q}_{\alpha_1...\alpha_n}{}^*$ and identify the equivalence classes $C_x(n)$. Since the subsystem $\mathbf{Q}_{\alpha_1...\alpha_n}{}^*$ may be characterized as a queueing network defined by $M(R\text{-}n,\mathbf{k}_R+...+\mathbf{k}_{R-n+1})$, the distribution $\mathbf{v}_{\alpha_1...\alpha_n}{}^*$ corresponds to that of the queueing network $M(R\text{-}n,\mathbf{k}_R+...+\mathbf{k}_{R-n+1})$. The state of this network is $\mathbf{n}^{(R-n)}$ and, using eq. 2.10, the state-distribution $\pi(\mathbf{n}^{(R-n)},\mathbf{k}_R+...+\mathbf{k}_{R-n+1})$ is seen to be

$$\pi(\mathbf{n}^{(R-n)},\mathbf{k}_R+...+\mathbf{k}_{R-n+1}) = G_{R-n}(\mathbf{k}_R+...+\mathbf{k}_{R-n+1})^{-1}\left\{\prod_{i=1}^{P}\left(n_i^{(R-n)}! \Big/ \prod_{a=1}^{n_i^{(R-n)}} \beta_i(a,k_{iR}+...+k_{i(R-n+1)})\right)\right\} \left\{\prod_{i=P+1}^{N}\prod_{r=1}^{R-n} w_{ir}{}^{n_{ir}}/n_{ir}!\right\}, \tag{4.18}$$

where nodes (P+1),...,N are assumed to be of the IS type,

$\beta_i(a,k_{iR}+...+k_{i(R-n+1)}) = a/(a+k_{iR}+...+k_{i(R-n+1)})$, $n_i^{(R-n)} = \sum_{r=1}^{R-n} n_{ir}$ and

$$G_{R-n}(\mathbf{k}_R+\ldots+\mathbf{k}_{R-n+1}) =$$

$$\sum_{\mathbf{n}^{(R-n)} \in S^{(R-n)}} \{\prod_{i=1}^{P} (n_i^{(R-n)}! / \prod_{a=1}^{n_i^{(R-n)}} \beta_i(a, k_{iR}+\ldots+k_{i(R-n+1)}))\} \{\prod_{i=P+1}^{N} \prod_{r=1}^{R-n} w_{ir}^{n_{ir}}/n_{ir}!\}.$$

We now identify the equivalence classes to be found at level n when the adopted state-space partitioning follows the Decomposition by Routing Chain method. Two subsystems $\mathbf{Q}_{\alpha_1\ldots\alpha_n}{}^*$ and $\mathbf{Q}_{\beta_1\ldots\beta_n}{}^*$, where $\alpha_1 = \zeta^{(1)}(\mathbf{k}_R)$, $\alpha_2 = \zeta^{(2)}(\mathbf{k}_{R-1})$, ..., $\alpha_n = \zeta^{(n)}(\mathbf{k}_{R-n+1})$, $\beta_1 = \zeta^{(1)}(\mathbf{b}_R)$, ..., $\beta_n = \zeta^{(n)}(\mathbf{b}_{R-n+1})$, belong to the same equivalence class $C_x(n)$ if $\mathbf{v}_{\alpha_1\ldots\alpha_n}{}^* = \mathbf{v}_{\beta_1\ldots\beta_n}{}^*$. This condition is seen to be satisfied if

$$(\mathbf{k}_R+\ldots+\mathbf{k}_{R-n+1}) = (\mathbf{b}_R+\ldots+\mathbf{b}_{R-n+1}).$$

In other words, the subsystems $\mathbf{Q}_{\alpha_1\ldots\alpha_n}{}^*$ and $\mathbf{Q}_{\beta_1\ldots\beta_n}{}^*$ are equivalent if the parameter vectors entering into the state-dependent service-rate functions $\beta_i(.)$ are the same. Since at the nth level in the hierarchical state-space partitioning $\mathbf{k}_r \in L_r$, for $R-n+1 \le r \le R$, where

$$L_r = \{\mathbf{k}_r \mid k_{ir} \ge 0,\ 1 \le i \le N,\ \sum_{i=1}^{N} k_{ir} = K_r,\ 1 \le r \le R\},$$

the number of distinct equivalence classes at level n is

$$E_n = \binom{K_R+\ldots+K_{R-n+1}+N-1}{N-1},$$

this being the cardinality of the set

$$\{\mathbf{k} \mid \mathbf{k} = \sum_{r=R-n+1}^{R} \mathbf{k}_r,\ \mathbf{k}_r \in L_r,\ R-n+1 \le r \le R\}.$$

At the first level, we have $E_1 = \Omega(1)$. However, for $n \geq 2$, the number of equivalence classes at level n is much smaller than the width $\Omega(n)$. The width is given explicitly by eq. 4.1 with

$$M = \binom{K_R+N-1}{N-1}, \quad \chi(\alpha_1) = \binom{K_{R-1}+N-1}{N-1}, \quad \chi(\alpha_1,\alpha_2) = \binom{K_{R-2}+N-1}{N-1}$$

and, in general,

$$\chi(\alpha_1,\ldots,\alpha_n) = \binom{K_{R-n}+N-1}{N-1}.$$

The third step in the construction is to establish a recursive relationship $R_x(n)$ between the system $\mathbf{Q}_x(n)$, corresponding to the equivalence class $C_x(n)$ at level n, and the set of subsystems $F_x(n)$. With the method of state-space partitioning that we have adopted, the system $\mathbf{Q}_x(n)$ is a queueing network denoted by $M(R\text{-}n,\mathbf{k}_R+\ldots+\mathbf{k}_{R-n+1})$ such that $x = \eta^{(R-n+1)}(\mathbf{k}_R+\ldots+\mathbf{k}_{R-n+1})$, where $\eta^{(R-n+1)}(.)$ is a function which maps the elements of the set $\{(\mathbf{k}_R+\ldots+\mathbf{k}_{R-n+1}) \mid \mathbf{k}_r \in L_r, (R\text{-}n+1) \leq r \leq R\}$ into the set of integers

$$\{1,2,\ldots,\binom{K_R+\ldots+K_{R-n+1}+N-1}{N-1}\}.$$

The state-distribution for this system is given by eq. 4.18. Now consider the state-distributions of the systems in the set $F_x(n)$. We have

$$F_x(n) = \{\mathbf{Q}_{\alpha_1\ldots\alpha_n\alpha_{n+1}}{}^* \mid \alpha_{n+1} = \zeta^{(n+1)}(\mathbf{k}_{R-n}), \mathbf{k}_{R-n} \in L_{R-n}\}.$$

Since the subsystem $\mathbf{Q}_{\alpha_1\ldots\alpha_{n+1}}{}^*$ may be characterized as a queueing network defined by $M(R\text{-}n\text{-}1,\mathbf{k}_R+\ldots+\mathbf{k}_{R-n})$, the state-distribution $\mathbf{v}_{\alpha_1\ldots\alpha_{n+1}}{}^*$ corresponds exactly to $\pi(\mathbf{n}^{(R-n-1)},\mathbf{k}_R+\ldots+\mathbf{k}_{R-n})$, as given by eq. 4.18, but with n replaced by (n+1).

We now relate the state-distribution $\pi(\mathbf{n}^{(R-n)},\mathbf{k}_R+...+\mathbf{k}_{R-n+1})$ to $\pi(\mathbf{n}^{(R-n-1)},\mathbf{k}_R+...+\mathbf{k}_{R-n})$. By definition,

$$\pi(\mathbf{n}^{(R-n)},\mathbf{k}_R+...+\mathbf{k}_{R-n+1}) =$$

$$\pi(\mathbf{n}^{(R-n-1)},\mathbf{k}_R+...+\mathbf{k}_{R-n}) \sum_{\substack{\mathbf{n}^{(R-n)} \in S^{(R-n)} \\ \mathbf{n}_{R-n} = \mathbf{k}_{R-n}}} \pi(\mathbf{n}^{(R-n)},\mathbf{k}_R+...+\mathbf{k}_{R-n+1}), \tag{4.19}$$

where $\mathbf{n}_{R-n} = (n_{1(R-n)},...,n_{N(R-n)})$. Using eq. 4.19, we now derive the relationship between the normalization constants associated with the system $M(R-n,\mathbf{k}_R+...+\mathbf{k}_{R-n+1})$, corresponding to $\mathbf{Q}_x(n)$, and the set of systems $F_x(n)$. Substituting eq. 4.18 into eq. 4.19, we obtain

$$\sum_{\substack{\mathbf{n}^{(R-n)} \in S^{(R-n)} \\ \mathbf{n}_{R-n} = \mathbf{k}_{R-n}}} \pi(\mathbf{n}^{(R-n)},\mathbf{k}_R+...+\mathbf{k}_{R-n+1}) =$$

$$g_{R-n}G_{R-n-1}(\mathbf{k}_R+...+\mathbf{k}_{R-n+1}+\mathbf{k}_{R-n}) \,/\, G_{R-n}(\mathbf{k}_R+...+\mathbf{k}_{R-n+1}),$$

where

$$g_{R-n} = \{\prod_{i=1}^{P} w_{i(R-n)}{}^{k_{i(R-n)}} \binom{k_{i(R-n)}+k_{iR}+...+k_{i(R-n+1)}}{k_{iR}+...+k_{i(R-n+1)}}\} \{\prod_{i=P+1}^{N} w_{i(R-n)}{}^{k_{i(R-n)}}/k_i{}^{(R-n)}!\ \}.$$

Summing both sides of the above over all $\mathbf{k}_{R-n} \in L_{R-n}$, we obtain

$$G_{R-n}(\mathbf{k}_R+...+\mathbf{k}_{R-n+1}) = \sum_{\mathbf{k}_{R-n} \in L_{R-n}} g_{R-n}G_{R-n-1}(\mathbf{k}_R+...+\mathbf{k}_{R-n}).$$

Making the transformation of variables $\mathbf{v}_r = (\mathbf{k}_R+...+\mathbf{k}_{R-n+1})$ and $r = (R-n)$, we obtain

$$G_r(\mathbf{v}_r) = \sum_{\mathbf{k}_r \in L_r} g_r G_{r-1}(\mathbf{v}_r+\mathbf{k}_r). \tag{4.20}$$

Equation 4.20 is the fundamental underlying recursive equation on which RECAL is based in the case of constant service-rate functions. It expresses the relationship between the normalization constant of the subsystem $M(R\text{-}n,\mathbf{k}_R+\ldots+\mathbf{k}_{R\text{-}n+1})$ at level n and those associated with the set of subsystems

$$\{M(R\text{-}n\text{-}1,\mathbf{k}_R+\ldots+\mathbf{k}_{R\text{-}n}) \mid \mathbf{k}_{R\text{-}n} \in L_{R\text{-}n}\}$$

at level (n+1) in the hierarchy.

In the case of state-dependent service-rate functions of the form $\mu_i(n_i)$, the relationship expressed by eq. 4.20 retains the same form but g_r is simply replaced by

$$g_r = \{\prod_{i=1}^{P} w_{ir}{}^{k_{ir}} \binom{k_{ir}+k_{iR}+\ldots+k_{i(r+1)}}{k_{iR}+\ldots+k_{i(r+1)}} / \prod_{a=1}^{k_{ir}} \mu_i(a+k_{iR}+\ldots+k_{i(r+1)})\}$$
$$\{\prod_{i=P+1}^{N} w_{ir}{}^{k_{ir}}/k_{ir}!\}.$$

We now consider the space and time requirements of the recursion. Since the number of distinct equivalence classes at level n in the hierarchy is

$$E_n = \binom{\sum_{r=R-n+1}^{R} K_r + N - 1}{N-1},$$

the space requirement of the recursion is of the order of

$$\binom{K+N-1}{N-1}, \qquad (4.21)$$

where $K = \sum_{r=1}^{R} K_r$ is the total population of customers in the network $B(N,\mathbf{K})$. The time requirement is of the order of

$$\sum_{n=1}^{R} \binom{\sum_{r=R-n+1}^{R} K_r + N - 1}{N-1}. \tag{4.22}$$

If we consider N as being a fixed quantity and assume, for the sake of simplicity, that all routing chains have the same population, then the growth of the space and time requirements, as given by eqs. 4.21 and 4.22, respectively, may be written as polynomials in the variable R. This factor gives RECAL a computational advantage over the Convolution Algorithm and MVA when R is large relative to N.

4.7 Constructive Derivation of MVAC

There are two versions of the MVAC algorithm that have been developed. The first, which is termed here as '*parallel*' MVAC, is applicable to closed BCMP queueing networks with state-dependent service-rate functions [CON2,CON9]. The term 'parallel' is used since, in this version, the mean performance measures for all of the routing chains in the network are obtained in a single pass through the recursive equations which are utilized. The Parallel MVAC algorithm involves the computation of certain marginal queue-length distributions. In the second version, termed '*sequential*' MVAC, the performance measures for the R routing chains are determined sequentially by making R passes through the recursion [CON6,SOU1]. There are two separate formulations of the Sequential MVAC algorithm. The first formulation is applicable to networks with constant service-rate functions and makes use of the MVA waiting-time equation, as given by eq. 4.16. The recursion utilized in this formulation is cast entirely in terms of the mean performance measures of certain networks. The second formulation is applicable to networks with state-dependent service-rate functions. The recursion utilized in this formulation involves the marginal queue-length distributions of certain networks. The second formulation of the Sequential MVAC algorithm is related directly to the Parallel MVAC algorithm. Both make use of a recursive equation that is in terms of marginal queue-length distributions.

In this Section, we shall only develop the recursive structure and the fundamental recursive equation for the case of the second formulation of the Sequential MVAC algorithm. The structure and the fundamental equation of the parallel MVAC algorithm are exactly the same as in this case. The structure of the first formulation of Sequential MVAC is closely related to that of the second formulation. The first formulation makes use of the same recursive structure as the second formulation, but use is also made of the MVA waiting-time equation. For the sake of brevity, in the following we shall simply refer to the second formulation of the Sequential MVAC algorithm as the MVAC algorithm.

The recursive structure and the underlying equations of the MVAC algorithm may be developed by following essentially the same steps that have been followed in the development of RECAL. Hence, in the developments to be made below, we shall only consider those steps that differ from the developments made in Section 4.6. The main difference is that, instead of establishing a relationship among the normalization constants of the system $M(R\text{-}x,\mathbf{k}_R+\ldots+\mathbf{k}_{R-x+1})$ and the set of systems $F_x(n)$, we establish a relationship among the marginal queue-length distributions. This results in an algorithm that does not involve the computation of normalization constants and which circumvents the possible problems of numerical stability commonly associated with computing these quantities [CON3,LAM1]. The Recursion by Chain Algorithm and the MVAC algorithm are very closely related to one another since both may be viewed as being based on exactly the same hierarchical method of state-space partitioning. As a result, both RECAL and MVAC have essentially the same computational costs. In the following, we again consider the BCMP queueing network denoted by $B(N,\mathbf{K})$ except that *we now assume that in each routing chain r there is only a single customer.* Hence, we assume that $K_r = 1$ for $1 \leq r \leq R$. This assumption, however, involves no loss of generality since it is not being assumed that each of these single-customer routing chains must have distinct service-requirements and routing-parameters. A routing chain r with population K_r may, equally well, be viewed as K_r routing chains with identical parameters and which each contain a single customer. The reason for adopting this assumption in the MVAC algorithm is that it results in a set of simpler recursive equations.

Under the assumption that $K_r = 1$ for $1 \le r \le R$, the system $M(R-n,\mathbf{k}_R+...+\mathbf{k}_{R-n+1})$ considered in Section 4.6 is a queueing network that contains the first (R-n) customers of $B(N,\mathbf{K})$. Since it is assumed that $K_r = 1$, the parameter vector $\mathbf{k}_r$, $R-n+1 \le r \le R$, is now an all-zero vector with a one in the position corresponding to the position of customer r in the network. Hence,

$$\mathcal{L}_r = \{\mathbf{k}_r \mid \sum_{i=1}^{N} k_{ir} = 1,\ k_{ir} = 0 \text{ or } 1\}.$$

The cardinality of the set $\mathcal{L}_r$ is, therefore, equal to N and the number of distinct equivalence classes at level n in the hierarchy is

$$E_n = \binom{n+N-1}{N-1}.$$

We now proceed to derive the relationship $R_x(n)$ in terms of marginal queue-length distributions. Consider the queueing network $M(R-n,\mathbf{k}_R+...+\mathbf{k}_{R-n+1})$. Let $P_i^{(R-n)}(x,\mathbf{k}_R+...+\mathbf{k}_{R-n+1})$ denote the probability that there are x customers, $0 \le x \le R-n$, at node i. As has been explained in Section 3.4, the queueing network $M(R-n,\mathbf{k}_R+...+\mathbf{k}_{R-n+1})$ may either be viewed as a network which consists of the first (R-n) chains of $B(N,\mathbf{K})$ and which contains state-dependent service-rate functions of the form

$$\beta_i(x,k_{iR}+...+k_{i(R-n+1)}) = x\mu_i(x+k_{iR}+...+k_{i(R-n+1)})/(x+k_{iR}+...+k_{i(R-n+1)})$$

or as a network at which there are $(k_{iR}+...+k_{i(R-n+1)})$ 'single-customer self-looping' (SCSL) chains at node i. Let us adopt the latter viewpoint. Consider now the network $M(R\text{-}n\text{-}1,\mathbf{k}_R+...+\mathbf{k}_{R-n+1}+\mathbf{1}_j)$. In this case, $\mathbf{k}_{R-n} = \mathbf{1}_j$ and the single customer of chain (R-n) is 'self-looping' at node j. Since the system $M(R\text{-}n\text{-}1,\mathbf{k}_R+...+\mathbf{k}_{R-n+1}+\mathbf{1}_j)$ is the system $M(R\text{-}n,\mathbf{k}_R+...+\mathbf{k}_{R-n+1})$ conditioned on the event that $n_{j(R-n)} = 1$, we may write

$$P_i^{(R-n)}(x,\mathbf{k}_R+...+\mathbf{k}_{R-n+1}) =$$

$$\pi(n_{i(R-n)} = 1)\ P_i^{(R-n-1)}(x-1,\mathbf{k}_R+...+\mathbf{k}_{R-n+1}+\mathbf{1}_i)$$

$$+ \sum_{\substack{j=1 \\ j\neq i}}^{N} P_i^{(R-n-1)}(x,\mathbf{k}_R+...+\mathbf{k}_{R-n+1}+\mathbf{1}_j)\pi(n_{j(R-n)} = 1), \qquad (4.23)$$

where

$$\pi(n_{j(R-n)} = 1) = \sum_{\substack{\mathbf{n}^{(R-n)} \in S^{(R-n)} \\ (n_{1(R-n)},...,n_{N(R-n)}) = \mathbf{1}_j}} \pi(\mathbf{n}^{(R-n)},\mathbf{k}_R+...+\mathbf{k}_{R-n+1}).$$

In words, $\pi(n_{j(R-n)} = 1)$ is simply the marginal probability that the single customer of routing chain (R-n) in $M(R\text{-}n,\mathbf{k}_R+...+\mathbf{k}_{R-n+1})$ is at node j.

We now find an expression for $\pi(n_{j(R-n)} = 1)$ in terms of the queue-length distributions $P_i^{(R-n-1)}(x,\mathbf{k}_R+...+\mathbf{k}_{R-n+1}+\mathbf{1}_j)$, $1 \le j \le N$. Given that the single customer of chain (R-n) is at node j in $M(R\text{-}n,\mathbf{k}_R+...+\mathbf{k}_{R-n+1})$, the probability flux $\tau_{R-n}(\mathbf{k}_R+...+\mathbf{k}_{R-n+1},j)$ at which this customer moves *out of* node j is

$$\tau_{R-n}(\mathbf{k}_R+...+\mathbf{k}_{R-n+1},j) =$$

$$t_{j(R-n)}^{-1} \sum_{x=0}^{R-n-1} P_j^{(R-n-1)}(x, \mathbf{k}_R + \ldots + \mathbf{k}_{R-n+1} + \mathbf{1}_j)$$

$$\mu_j(x + k_{jR} + \ldots + k_{j(R-n+1)} + 1)/(x + k_{jR} + \ldots + k_{j(R-n+1)} + 1).$$

Since the actual proportion of time that the customer of chain (R-n) spends at node j is proportional to $e_{j(R-n)}\tau_{R-n}^{-1}(\mathbf{k}_R + \ldots + \mathbf{k}_{R-n+1}, j)$, we may write

$$\pi(n_{j(R-n)} = 1) =$$
$$e_{j(R-n)}\tau_{R-n}^{-1}(\mathbf{k}_R + \ldots + \mathbf{k}_{R-n+1}, j) / \sum_{i=1}^{N} e_{i(R-n)}\tau_{R-n}^{-1}(\mathbf{k}_R + \ldots + \mathbf{k}_{R-n+1}, i) \quad (4.24)$$

Applying the transformation of variables $r = (R-n)$ and $\mathbf{v}_r = \mathbf{k}_R + \ldots + \mathbf{k}_{R-n+1}$ to eqs. 4.23 and 4.24, we obtain the more concise expression

$$P_i^{(r)}(x, \mathbf{v}_r) = \pi(n_{ir} = 1)P_i^{(r-1)}(x-1, \mathbf{v}_r + \mathbf{1}_i) + \sum_{\substack{j=1 \\ j \neq i}}^{N} P_i^{(r-1)}(x, \mathbf{v}_r + \mathbf{1}_j)\pi(n_{jr} = 1), \quad (4.25)$$

where

$$\pi(n_{jr} = 1) = e_{jr}\tau_r^{-1}(\mathbf{v}_r, j) / \sum_{i=1}^{N} e_{ir}\tau_r^{-1}(\mathbf{v}_r, i). \quad (4.26)$$

Equations 4.25 and 4.26 are the fundamental recursive equations on which the Parallel MVAC algorithm and the second formulation of the Sequential MVAC algorithm are based. The same equations may, of course, be used when we have constant service-rate functions.

4.8 Constructive Derivation of DAC

In this Section, we consider the derivation of the recursive structure and the fundamental recursion of the Distribution Analysis by Chain (DAC) algorithm. In contrast to the other algorithms that we have considered in the previous four sections that are primarily intended for the computation of the mean performance measures of multiple-chain queueing networks, the DAC algorithm is expressly intended for the computation of the joint marginal queue-length distribution. One application in which this distribution is of importance is in the evaluation of computer system availability models, as previously described in Subsection 2.3.8.

The construction of the recursive structure of DAC and the determination of the main recursive equation which is utilized may be carried out by following essentially the same steps that were followed in the construction of MVA, as carried out in Section 4.5. As in Section 4.7, we shall assume in the following, without loss of generality, that $K_r = 1$ for $1 \leq r \leq R$. We shall use the notation $B_r(N,\mathbf{K}_r)$, where $\mathbf{K}_r = (K_1,\ldots,K_r)$ and $K_s = 1$ for $1 \leq s \leq r$, to denote a queueing network that is identical to $B(N,\mathbf{K})$, but with the customers of the last (R-r) (single-customer) routing chains removed from the network. With this notation, the network $B(N,\mathbf{K})$ is denoted by $B_R(N,\mathbf{K}_R)$. In the following, we shall assume that we have state-dependent service-rate functions $\mu_i(.)$ at the nodes. We shall also assume that there are no IS centers in the network ($P = N$). If there are IS queues, then these are treated as single-server queues with $\mu_i(x) = x$.

Consider the queueing network $B_R(N,\mathbf{K}_R)$ with state-space $S^{(R)}$ and suppose that we partition $S^{(R)}$ according to the disposition in the network of the single customer of chain R. This method of single-level partitioning is similar to that adopted in the construction of MVA algorithm. Let $P^R(\mathbf{k})$, where $\mathbf{k} = (k_1,\ldots,k_N)$, denote the probability that ($n_1 = k_1$, ..., $n_N = k_N$) in the network $B_R(N,\mathbf{K}_R)$, where n_i is the total number of customers at node i. Using eq. 2.10, we have by definition

$$P^R(\mathbf{k}) = \sum_{\substack{\mathbf{n}^{(R)} \in S^{(R)} \\ \sum_{r=1}^{R} n_{ir} = k_i \\ 1 \le i \le N}} G(\mathbf{K}_R)^{-1}\{\prod_{i=1}^{N} (n_i!/\prod_{a=1}^{n_i} \mu_i(a)) \prod_{r=1}^{R} w_{ir}^{n_{ir}}/n_{ir}!\},$$

where $G(\mathbf{K}_R)$ is the normalization constant associated with the state-distribution of $B_R(N,\mathbf{K}_R)$. We now consider the relationship between the system $\mathbf{Q}_1(0)$, corresponding to $B_R(N,\mathbf{K}_R)$ and the subsystems $\mathbf{Q}_{\alpha_1}(1)$ corresponding to $\mathbf{Q}_{\alpha_1}{}^*$, where $1 \le \alpha_1 \le N$. By definition

$$P^R(\mathbf{k},n_{jR} = 1) = P^R(\mathbf{k} \mid n_{jR} = 1)u_j, \tag{4.27}$$

where u_j is the probability that $n_{jR} = 1$ in the network $B_R(N,\mathbf{K}_R)$. By definition,

$$P^R(\mathbf{k} \mid n_{jR} = 1) =$$

$$\sum_{\substack{\mathbf{n}^{(R-1)} \in S^{(R-1)} \\ n_i^{(R-1)} = k_i,\ 1 \le i \le N,\ i \ne j \\ n_j^{(R-1)} = k_j - 1}} G(\mathbf{K}_R)^{-1} w_{jR}((n_j^{(R-1)}+1)/\mu_j(n_j^{(R-1)}+1))$$

$$\{\prod_{i=1}^{N} (n_i^{(R-1)}!/\prod_{a=1}^{n_i^{(R-1)}} \mu_i(a)) \prod_{r=1}^{R-1} w_{ir}^{n_{ir}}/n_{ir}!\}u_j^{-1}, \tag{4.28}$$

where $n_i^{(R-1)} = \sum_{r=1}^{R-1} n_{ir}$. Hence, substituting eq. 4.28 into eq. 4.27, we have

$$P^R(\mathbf{k},n_{jR} = 1) = k_j w_{jR} G(\mathbf{K}_{R-1}) P^{R-1}(\mathbf{k}-\mathbf{1}_j) / \mu_j(k_j) G(\mathbf{K}_R).$$

In terms of the notation adopted here we have [BRU1]

$$T_{jR}(\mathbf{K}_R) = e_{jR} G(\mathbf{K}_{R-1})/G(\mathbf{K}_R)$$

so that

$$P^R(\mathbf{k}, n_{jR} = 1) = k_j t_{jR} T_{jR}(\mathbf{K}_R) P^{R-1}(\mathbf{k}-\mathbf{1}_j)/\mu_j(k_j). \tag{4.29}$$

Summing both sides of eq. 4.29 over all j, $1 \leq j \leq N$, we obtain finally

$$P^R(\mathbf{k}) = \sum_{j=1}^{N} k_j t_{jR} T_{jR}(\mathbf{K}_R) P^{R-1}(\mathbf{k}-\mathbf{1}_j)/\mu_j(k_j) \tag{4.30}$$

which is the fundamental recursive equation of the DAC algorithm. Hence, by applying a single-level decomposition based on the position of the single customer of chain R in $B_R(N,\mathbf{K}_R)$, we obtain a relationship between the joint marginal queue-length distribution of the system $B_R(N,\mathbf{K}_R)$ and that associated with $B_{R-1}(N,\mathbf{K}_{R-1})$.

If we continue to apply the above described single-level decomposition to $B_{R-1}(N,\mathbf{K}_{R-1})$, then at the *r*th level in the resulting hierarchy we have the system $B_{R-r}(N,\mathbf{K}_{R-r})$. Consequently, the number of levels in the hierarchy is R and the width of each level in *one*. At level r, we have the joint marginal queue-length distribution

$$\{P^{R-r}(\mathbf{k}) \mid \mathbf{k} = (k_1,\ldots,k_N),\ k_i \geq 0,\ \sum_{i=1}^{N} k_i = R-r\}$$

associated with $B_{R-r}(N,\mathbf{K}_{R-r})$. These considerations imply that the computational and storage requirements of DAC are proportional to the cardinality of the set

$$\{P^{R}(\mathbf{k}) \mid \mathbf{k} = (k_1,\ldots,k_N),\ k_i \geq 0,\ \sum_{i=1}^{N} k_i = R\}$$

which is $\binom{K+N-1}{N-1}$, where K (K = R) is the total population of $B_R(N,\mathbf{K}_R)$. Hence, the DAC algorithm may be considered to be an optimal algorithm for computing the joint marginal queue-length distribution of $B(N,\mathbf{K})$.

4.9 Hybrid Decompositions and Hybrid Algorithms

In the previous five sections, we have seen how the main exact algorithms that have hitherto been developed may be constructed according to the methodology that we have developed in Section 4.3. We showed how the Convolution Algorithm could be developed starting from a hierarchical state-space partitioning that is conditioned on the population vector of particular nodes in the network. The MVA algorithm and the DAC algorithm were developed starting from a single-level partitioning conditioned on the position in the network of a particular customer. The Recursion by Chain Algorithm and MVAC were developed starting from a hierarchical state-space partitioning conditioned on the state-vector of customers of particular routing chains. The general methodology that we have proposed, however, suggests the existence of other algorithms that could possibly be developed based on other methods of state-space partitioning. In view of the fact that we may arbitrarily decompose a reversible queueing network and maintain exact results by the general Decomposition and Aggregation procedure, it is quite natural to give consideration to other methods of state-space partitioning and attempt to develop new algorithms following the methodology that we have formulated.

One particular alternate method of state-space partitioning that is perhaps apparent at this point is a hierarchical hybrid decomposition in which we begin, at the first level, by decomposing the state-space $S^{(R)}$ of $B(N,\mathbf{K})$ according to the Decomposition by Routing Chain method and then switch to the method of Decomposition by Service Center, after a certain level in the hierarchy has been reached. Alternatively, we may consider beginning with the Decomposition by Service Center method and then switching to the Decomposition by Routing Chain method. An algorithm based on the former hybrid method of decomposition has been suggested as an efficient means to compute the normalization constant associated with the queueing network $B(N,\mathbf{K})$ [CON5]. If we construct an algorithm involving normalization constants based on such a hybrid decomposition then, by varying the 'switch-over point', we will have at one extreme the Recursion by Chain Algorithm and at the other extreme the Convolution Algorithm. If the switch-over point is made at some intermediate level, then, by its very generality, the resulting hybrid algorithm may be more efficient computationally than either RECAL or the Convolution Algorithm. Such hybrid algorithms, however, have not been studied yet in any detail.

It is not our purpose here to develop hybrid computational algorithms in further detail. Rather, we merely cite these possibilities to illustrate the potential offered by the general methodology that we have formulated. The development of hybrid algorithms, based on hybrid methods of state-space partitioning, and the development of other algorithms based on alternate methods of partitioning, appears, however, to be an interesting direction for future research.

4.10 In Search of the Optimal Exact Computational Algorithm

A provoking question which is raised by the general constructive theory that we have proposed for exact computational algorithms concerns the form of the algorithm which is computationally optimal within the class of possible algorithms defined by our methodology. Given a particular closed BCMP queueing network B(N,**K**) with state-space $S^{(R)}$, the general question that may be posed is how we should partition $S^{(R)}$ to arrive at an algorithm to compute mean performance measures with computational requirements that are minimized. This question, which concerns the inherent computational complexity of queueing network problems, is one that remains open. The great difference between the computational growth characteristics of the Convolution Algorithm and MVA, on the one hand, and RECAL and MVAC, on the other, makes this question perhaps even more interesting. Formally, we may pose the question of constructing the optimal computational algorithm in terms of the following nonlinear optimization problem:

Given the BCMP queueing network $B(N,\mathbf{K})$ *with state-space* $S^{(R)}$, *find the sets*

$$\{ S(\alpha_1) \mid 1 \leq \alpha_1 \leq M, S^{(R)} \supset S(\alpha_1)\}$$

and the sets

$$\{S(\alpha_1,...,\alpha_n) \mid 1 \leq \alpha_1 \leq M, 1 \leq \alpha_i \leq \chi(\alpha_1,...,\alpha_{i-1}), i = 2,...,n, 2 \leq n \leq L-1\}$$

so that

$$\Lambda(\text{Max} \{ E_n \mid 1 \leq n \leq L\}, \sum_{n=0}^{L} E_n)$$

is minimized, where E_n *is the number of distinct equivalence classes at level* n *in the hierarchy,* L *is the total number of levels in the hierarchical state-space decomposition,* $\chi(\alpha_1,...,\alpha_i)$ *is the number of sets of communicating states into which the set of states* $S(\alpha_1,...,\alpha_i)$ *is partitioned and, finally,* $\Lambda(.)$ *is some cost function involving the space and time requirements of the algorithm which results from the method of state-space partitioning that is adopted.*

The optimization problem described above is, of course, a complex one. Since the state-space $S^{(R)}$ depends directly on the number of nodes N, the number of closed routing chains R, the chain populations K_r and also on the visit-ratios e_{ir} if any of these happen to be equal to zero, it is apparent that the optimal algorithm will take on different forms depending on the actual values of these parameters. Hence, we may conceivably view the optimal algorithm as an adaptive algorithm involving two main steps. In the first step, we determine the optimal form and then, in the second step, we actually solve the queueing network under consideration for the mean performance measures. In theory, the computational costs involved in the first step should also be taken into account in the overall optimization.

In view of the above speculation, it appears reasonable to conjecture on the existence of an exact computational algorithm for mean performance measures that is markedly different in structure from the algorithms that have been developed and which may offer the possibility of significant improvements in efficiency relative to the Convolution Algorithm, MVA, RECAL, MVAC or DAC. The constructive methodology that we have proposed, based on probabilistic concepts, provides much intuitive insight and may possibly lead to progress in this direction.

CHAPTER 5

Exact Computational Algorithms for Queueing Networks

The main general purpose exact computational algorithms that have been developed for BCMP queueing networks include the Convolution Algorithm, Mean Value Analysis (MVA), the Recursion by Chain Algorithm (RECAL), Mean Value Analysis by Chain (MVAC) and the Distribution Analysis by Chain (DAC) algorithm. Certain other closely related variations have also been developed, including the Local Balance Algorithm for Normalizing Constants (LBANC) [CHA3], Hybrid Convolution-LBANC [LAV2] and the Normalized Convolution Algorithm (NCA) [REI9]. In the previous chapter, we showed how the recursive structures and the fundamental underlying recursive equations of the main algorithms could be derived within the framework of a general methodology based on the notions of Decomposition and Aggregation. In this final chapter, we provide a survey of the details of these algorithms.

In the following, we limit our attention to the solution of closed, multiple-chain BCMP queueing networks and mixed, multiple-chain BCMP queueing networks with 'limited queue-dependent' service-rate functions (to be defined). These are commonly the cases of greatest practical interest. Exact algorithms have also been developed for queueing networks with population size constraints [LAM4,GEO2], generalized state-dependent routing [KRZ1,KRZ2,SAU1,TOW1,TOW2] and for BCMP networks that contain queues having the MSCCC service discipline [LEB2]. It is not our objective, however, to consider the solution of these more general types of product-form queueing networks here. We also do not consider the case of mixed networks with state-dependent arrival rates or networks with service-rate functions of the form $\mu_{ir}(\mathbf{n}_i^{(R)})$ [BAL1,SAU1]. The more general algorithms that have been developed for these more general situations are, however, closely related to those that apply to the basic closed BCMP queueing network that we shall consider here. As we have mentioned in Chapter 1, this chapter and Chapter 2 may be read together on their own without reference to the developments that have been made in Chapters 3 and 4. Much of the notation employed in this chapter has been defined previously in Subsection 2.1.3.

5.1 The Convolution Algorithm

The *Convolution Algorithm* is a so-called 'normalization constant approach' for obtaining the mean performance measures of product-form queueing networks since it involves the computation of the normalization constant which is associated with the state-distribution of the queueing network under consideration. The mean performance measures of interest are obtained directly in terms of this normalization constant and certain quantities that have been obtained in the course of computing it.

Consider, initially, a closed, multiple-chain BCMP queueing network B(N,**K**) with R closed routing chains, N nodes and population vector **K**, of the type considered in Subsection 2.1.3. Let the normalization constant associated with this queueing network be denoted by $C_N(\mathbf{K})$. Formally,

$$C_N(\mathbf{K}) = \sum_{\mathbf{n}^{(R)} \in S^{(R)}} \prod_{i=1}^{N} \left(f_i(\mathbf{n}_i^{(R)}) / \prod_{a=1}^{n_i} \mu_i(a)\right),$$

where the function $f_i(.)$ is as defined in Subsection 2.1.3. It may be shown [BRU1] that

$$C_N(\mathbf{K}) = \sum_{n_R=0}^{K_R} \cdots \sum_{n_2=0}^{K_2} \sum_{n_1=0}^{K_1} f_N(\mathbf{n}) C_{N-1}(\mathbf{K}-\mathbf{n}), \tag{5.1}$$

where $\mathbf{n} = (n_1,\ldots,n_R)$ and $C_{N-1}(\mathbf{K}-\mathbf{n})$ is the normalization constant associated with the queueing network $B(N-1,\mathbf{K}-\mathbf{n})$. We may write eq. 5.1 in the more concise form

$$C_N(\mathbf{K}) = (C_{N-1} \otimes f_N)(\mathbf{K}),$$

where $\otimes$ denotes an R-dimensional convolution operation. It follows from eq. 5.1 that

$$C_N(\mathbf{K}) = (f_1 \otimes \ldots \otimes f_N)(\mathbf{K}).$$

In the case of state-dependent service-rate functions of the form $\mu_i(n_i)$, where $n_i = \sum_{r=1}^{R} n_{ir}$, the Convolution Algorithm consists essentially of computing $C_N(\mathbf{K})$ using the recursive formula

$$C_m(\mathbf{k}) = (C_{m-1} \otimes f_m)(\mathbf{k}) \tag{5.2}$$

with the initial conditions $C_1(\mathbf{k}) = f_1(\mathbf{k})$. In the case that node m has a constant service-rate function, eq. 5.2 may be simplified and $C_m(\mathbf{k})$ may be computed using the simpler recursive formula [BRU1]

$$C_m(\mathbf{k}) = C_{m-1}(\mathbf{k}) + \sum_{r=1}^{R} w_{mr} C_m(\mathbf{k}-\mathbf{1}_r), \tag{5.3}$$

where $C_m(\mathbf{k}-\mathbf{1}_r) = 0$, if $k_r = 0$, $\mathbf{1}_r$ is a unit vector pointing in the direction r, $w_{mr} = t_{mr}e_{mr}$, and where the quantities t_{mr} and e_{mr} have been defined previously in Subsection 2.1.3.

Ignoring the computation of the initial conditions and assuming all nodes have constant service-rate functions, the total number of operations (multiplications and additions) to compute $C_N(\mathbf{K})$, using eq. 5.3, is

$$2R(N-1)\prod_{r=1}^{R}(K_r+1). \tag{5.4}$$

The storage space requirement, in number of array locations, is [BRU1]

$$2\prod_{r=1}^{R}(K_r+1). \tag{5.5}$$

If node m has a state-dependent service-rate function, then the computation of $C_m(\mathbf{k})$, for $\mathbf{0} \le \mathbf{k} \le \mathbf{K}$, using the previously computed values of $C_{m-1}(\mathbf{k})$, $\mathbf{0} \le \mathbf{k} \le \mathbf{K}$, requires of the order of (see for example [LAM3])

$$\prod_{r=1}^{R}(K_r+1)(K_r+2)/2$$

operations. The space requirement is the same as eq. 5.5.

We now consider how the mean performance measures may be calculated from the normalization constants obtained in the course of computing $C_N(\mathbf{K})$. We shall first consider the case where we have constant service-rate functions. We denote the throughput, utilization, mean queue-length (node population) and mean waiting-time (queueing and service-time) of type r customers at node i by T_{ir}, U_{ir}, Q_{ir} and W_{ir}, respectively. Also let $T_{ir}^{(c)}$, $U_{ir}^{(c)}$, $Q_{ir}^{(c)}$, and $W_{ir}^{(c)}$, denote the same respective quantities but on a per-class basis when we have class-switching in the network. As has been explained in Subsection 2.1.3, a BCMP queueing network with class-switching may be mapped, almost trivially, into a network with no class-switching.

When there are constant service-rate functions, the mean performance measures may be calculated as follows [LAV2]:

$$T_{ir} = e_{ir}C_N(\mathbf{K}-\mathbf{1}_r)/C_N(\mathbf{K}),$$

$$U_{ir} = t_{ir}T_{ir},$$

$$Q_{ir} = \sum_{a=1}^{K_r} w_{ir}{}^a C_N(\mathbf{K}-a\mathbf{1}_r)/C_N(\mathbf{K}),$$

$$W_{ir} = Q_{ir}/T_{ir},$$

$$T_{ir}{}^{(c)} = \alpha_{ic}{}^{(r)}T_{ir}/e_{ir},$$

$$U_{ir}{}^{(c)} = m_{ic}T_{ir}{}^{(c)},$$

$$Q_{ir}{}^{(c)} = \alpha_{ic}{}^{(r)}m_{ic}Q_{ir}/w_{ir},$$

$$W_{ir}{}^{(c)} = Q_{ir}{}^{(c)}/T_{ir}{}^{(c)}, \qquad (5.6)$$

where the quantities $\alpha_{ic}{}^{(r)}$ and m_{ic} have been defined previously in Subsection 2.1.3.

In the case of state-dependent service-rate functions of the form $\mu_i(n_i)$, the equations for T_{ir}, W_{ir}, $T_{ir}{}^{(c)}$, $Q_{ir}{}^{(c)}$ and $W_{ir}{}^{(c)}$ remain the same as in eqs. 5.6 but Q_{ir} is computed using [LAV2]

$$Q_{ir} = \sum_{x=1}^{K_r} x\ \pi(n_{ir}=x)$$

where

$$\pi(n_{ir}=x) = \sum_{\mathbf{k}_i \in A_{ir}(x)} \pi(\mathbf{n}_i{}^{(R)}=\mathbf{k}_i),$$

$$A_{ir}(x) = \{\mathbf{k}_i \mid \mathbf{k}_i = (k_{i1},\ldots,k_{iR}),\ 0 \le k_{is} \le K_s \text{ for } 1 \le s \le R,\ s \ne r,\ k_{ir} = x\},$$

$$\pi(\mathbf{n}_i{}^{(R)}=\mathbf{k}_i) = f_i(\mathbf{k}_i)C_{N-\{i\}}(\mathbf{K}-\mathbf{k}_i)/C_N(\mathbf{K})$$

and $C_{N-\{i\}}(\mathbf{k})$ is the normalization constant of a network which is identical to $B(N,\mathbf{k})$ but with the *i*th node removed, that is

$$C_{N-\{i\}}(\mathbf{k}) = (f_1 \otimes \ldots \otimes f_{i-1} \otimes f_{i+1} \otimes \ldots \otimes f_N)(\mathbf{k}).$$

A practical problem with the Convolution Algorithm is that the floating-point range of a machine may be exceeded in the course of computing $C_N(\mathbf{K})$. A partial solution is to scale the quantities $(e_{1r},\ldots,e_{Nr})$ by a constant factor for each chain r. This is termed *static scaling*. A scaling procedure is given in [LAV2, Section 3.5.4.b]. Another approach is to vary the scaling factors as we compute $C_N(\mathbf{K})$. Such a *dynamic scaling* procedure has been developed by Lam [LAM1].

We now consider how the mean performance measures may be obtained using the Convolution Algorithm when we have both open and closed routing chains in the network. In the following, we limit our attention to the case of a mixed network that contains a single open chain and limited queue-dependent service-rate functions. A *limited* queue-dependent service-rate function is one in which $\mu_i(n) = \mu_i(g_i)$ for $n \geq g_i$, where g_i, $g_i \geq 2$, is some integer constant [LAV2]. There is no loss in generality in limiting our attention to the case of a *single* open chain since a mixed network with several open chains and constant exogenous arrival rates is equivalent to a mixed network with a single open chain [SAU1]. The reason for limiting attention to the case of limited queue-dependent service-rate functions is that for general service-rate functions it may be difficult to establish if the mixed network is actually stable in the sense of the existence of a stationary distribution. With limited queue-dependent centers, a mixed network is stable if and only if [LAV2]

$$w_{ir_0}/\mu_i(g_i) < 1 \text{ for } 1 \leq i \leq N,$$

where r_0 is the chain index of the open chain.

The approach which is usually taken to analyze a mixed queueing network with a single open chain is first to transform it into an equivalent closed queueing network which has the same mean performance measures for the closed chains as in the original mixed network [SAU1]. The equivalent closed network is constructed from the original mixed network by removing the open chain and introducing special modified service-rate functions $\mu_i^*(n)$ at the nodes of the network. Having analyzed the equivalent closed network for the mean performance measures of the closed chains, the measures for the open chain may then be obtained.

The modified service-rate functions are given by [SAU2]

$$\mu_i^*(n) = \mu_i(n)\alpha_i(n-1)/\alpha_i(n),$$

where

$$\alpha_i(n) = \mu_i(g_i)^{g_i-n-1}/([\prod_{h=n+1}^{g_i-1} \mu_i(h)][(1-w_{ir_0}/\mu_i(g_i))^{n+1}])$$

$$+\sum_{a=1}^{g_i-2} \binom{n+a}{a} w_{ir_0}{}^a[\ 1/\prod_{h=n+1}^{n+a} \mu_i(h) - 1/\prod_{h=n+1}^{g_i-1} \mu_i(h)(\mu_i(g_i))^{n-g_i+1}\ ],$$

if $n \leq (g_i-2)$, and

$$\alpha_i(n) = 1/(1-w_{ir_0}/\mu_i(g_i))^{n+1},$$

if $n \geq (g_i-1)$. For constant service-rate functions, we have, therefore,

$$\mu_i^*(n) = (1-w_{ir_0}).$$

This implies that, for the case of constant service-rate functions, in the transformation from the mixed network to the equivalent closed network, we need merely replace w_{ir} by $w_{ir}/(1-w_{ir_0})$.

Having constructed the equivalent closed network with the modified service-rate functions $\mu_i^*(n)$, the mean performance measures for the closed chains are obtained directly by analyzing the closed network using the version of the Convolution Algorithm that applies to closed networks with state-dependent service-rate functions. Having analyzed the closed network for the mean performance measures of the closed routing chains, we may then find the performance measures for the open chain as follows [LAV2]:

If node i is limited queue-dependent, then

$$Q_{ir_0} = w_{ir_0} \sum_{k=1}^{K} (k+1)P_i(k,\mathbf{K})/\mu_i^*(k+1),$$

where $K = \sum_{r=1}^{R} K_r$, $P_i(x,\mathbf{K})$ is the probability that the total number of customers at node i in the equivalent closed network is x and

$$P_i(x,\mathbf{K}) = \sum_{\mathbf{k}_i \in B_i(x)} \pi(\mathbf{n}_i^{(R)}=\mathbf{k}_i),$$

where

$$B_i(x) = \{\ \mathbf{k}_i \mid \mathbf{k}_i = (k_{i1},\ldots,k_{iR}),\ 0 \le k_i \le K_s \text{ for } 1 \le s \le R,\ \sum_{s=1}^{R} k_{is} = x\}$$

and $\pi(\mathbf{n}_i^{(R)}=\mathbf{k}_i) = f_i(\mathbf{k}_i)C_{N-\{i\}}(\mathbf{K}-\mathbf{k}_i)/C_N(\mathbf{K})$. If node i has a constant service-rate function, then

$$Q_{ir_0} = (w_{ir_0}/(1-w_{ir_0}))(1 + \sum_{r=1}^{R} Q_{ir}).$$

If node i is IS, then

$$Q_{ir_0} = w_{ir_0}.$$

Finally,

$T_{ir_0} = e_{ir_0}$ and $W_{ir_0} = Q_{ir_0}/T_{ir_0}$.

5.2 MVA - Mean Value Analysis

The *Mean Value Analysis* (MVA) algorithm, in constrast with the Convolution Algorithm, involves a recursion that is cast directly in terms of mean performance measures and, in certain situations, marginal queue-length distributions. As a consequence, it avoids the potential problems of numerical stability associated with the Convolution Algorithm. The MVA algorithm was originally developed on the basis of the Arrival Theorem, as stated informally in Section 4.5, and Little's result [LIT1].

Let $P_i(j,\mathbf{k})$ denote the probability that there are j customers at node i in the closed, multiple-chain BCMP queueing network $B(N,\mathbf{k})$ with population vector $\mathbf{k}$, $\mathbf{k} = (k_1,\ldots,k_R)$. Also let $T_{ir}(\mathbf{k})$, $Q_{ir}(\mathbf{k})$ and $W_{ir}(\mathbf{k})$ denote the mean performance measures associated with the network $B(N,\mathbf{k})$. The following set of recursive equations constitute the MVA algorithm [BRU1,LAV2,REI4]:

$$W_{ir}(\mathbf{k}) = \begin{cases} t_{ir}, \text{ if node i is IS,} \\ t_{ir}[1+\sum_{s=1}^{R} Q_{is}(\mathbf{k}-\mathbf{1}_r)], \text{ if 'i' has a constant service-rate,} \\ t_{ir}\sum_{j=1}^{k} jP_i(j-1,\mathbf{k}-\mathbf{1}_r)]/\mu_i(j), \text{ if 'i' is state-dependent,} \end{cases}$$

$$T_{ir}(\mathbf{k}) = e_{ir}k_r/\sum_{j=1}^{N} e_{jr}W_{jr}(\mathbf{k}),$$

$$Q_{ir}(\mathbf{k}) = T_{ir}(\mathbf{k})W_{ir}(\mathbf{k}) \ ,$$

$$P_i(j,\mathbf{k}) = \sum_{s=1}^{R} t_i T_i(\mathbf{k}) P_i(j-1,\mathbf{k}-\mathbf{1}_s)/\mu_i(j),$$

$$P_i(0,\mathbf{k}) = 1 - \sum_{j=1}^{k} P_i(j,\mathbf{k}), \tag{5.7}$$

where $i = 1,...,N$, $r = 1,...,R$, $j = 1,...,k$, and $k = \sum_{s=1}^{R} k_s$.

The MVA algorithm consists of computing $W_{ir} = W_{ir}(\mathbf{K})$, $T_{ir} = T_{ir}(\mathbf{K})$ and $Q_{ir} = Q_{ir}(\mathbf{K})$, using eqs. 5.7 starting with the initial conditions $Q_{is}(\mathbf{0}) = 0$, $P_i(j,\mathbf{0}) = 0$, for $j > 0$, and $P_i(0,\mathbf{0}) = 1$. The recursion is over the population vector $\mathbf{k}$, where $\mathbf{0} \leq \mathbf{k} \leq \mathbf{K}$. When node i has a constant service-rate function, we also have

$$U_{ir} = t_{ir} T_{ir}(\mathbf{K}).$$

The mean performance measures, $T_{ir}^{(c)}$, $Q_{ir}^{(c)}$ and $W_{ir}^{(c)}$, on a per-class basis, may be obtained using eqs. 5.6. The equation for $U_{ir}^{(c)}$, appearing in eqs. 5.6, applies at a node with a constant service-rate function. If there are no state-dependent service-rate functions in the network, then the MVA algorithm need not involve the computation of the marginal queue-length distributions $P_i(j,\mathbf{k})$.

In the case when there are no state-dependent service-rate functions in the network, the time requirement, in number of operations, to obtain the mean performance measures, is approximately the same as eq. 5.4. The space requirement, in number of array locations, is [BRU1]

$$N \prod_{r=1}^{R} (K_r+1).$$

This is more than for the Convolution Algorithm when $N > 2$.

When there are state-dependent service-rate functions, the MVA algorithm is complicated by the need to compute and store the queue-length distributions $P_i(j,\mathbf{k})$. The algorithm may also fail numerically as $P_i(0,\mathbf{k})$ tends to zero. Reiser [REI9] has devised a means to circumvent this potential problem but it involves an increase in computational costs.

The analysis of mixed queueing networks, using the MVA algorithm, may be carried out in the same way as for the Convolution Algorithm by considering an equivalent closed queueing network, as previously described in Section 5.1. A comprehensive version of the MVA algorithm for mixed, multiple-chain BCMP queueing networks with general state-dependent service-rate functions may be found in [BRU2].

5.3 RECAL - The Recursion by Chain Algorithm

Like the Convolution Algorithm, the *Recursion by Chain Algorithm* (RECAL) is a 'normalization constant approach' for obtaining the mean performance measures of product-form queueing networks. The underlying recursion employed in RECAL is, however, markedly different from that employed in the Convolution Algorithm or MVA. The storage requirements and the computational costs are also very different. The time and space requirements of RECAL grow according to a polynomial function in the number of closed routing chains R, with the degree of the polynomial growing as a linear function of the number of nodes N. As a consequence, RECAL is substantially more efficient than the Convolution Algorithm or MVA when R is large relative to N. In [CON3], it has been established theoretically that RECAL begins to offer a significant computational advantage when R is approximately twice as large as N.

One particular application in which RECAL is of practical use is in the analysis of queueing network models of computer systems that have a large number of heterogeneous types of jobs. Another application is in the analysis of queueing models for OSI communication architectures, as described previously in Subsection 2.3.9, where we have two queues and one closed routing chain associated with each open entity in the seven layers of the OSI Reference Model. Yet another application is in the analysis of product-form system availability models, as described previously in Subsection 2.3.8, where we may have many types of components that are subject to failure and which have different failure/repair characteristics [CON6,GOY1,SOU1].

Consider the queueing network $B(N,\mathbf{K})$ and suppose, initially, that we have constant service-rate functions at the nodes. In contrast to the Convolution Algorithm whose recursion involves the normalization constants of queueing networks of the type defined by $B(n,\mathbf{k})$, where $\mathbf{0} \leq \mathbf{k} \leq \mathbf{K}$ and $1 \leq n \leq N$, the recursion employed in RECAL is in terms of the normalization constants of queueing networks denoted by $M(r,\mathbf{v}_r)$, where $\mathbf{v}_r = (v_{1r},\ldots,v_{Nr})$ and $M(r,\mathbf{v}_r)$ is a queueing network that contains the *first* r routing chains of $B(N,\mathbf{K})$ and in which special state-dependent service-rate functions $\beta_i(.)$ are introduced at the nodes of the network. The special form of these functions is

$$\beta_i(x,v_{ir}) = x/(x+v_{ir}),$$

where v_{ir} is a nonnegative integer to be specified.

Now let the normalization constant associated with the state-distribution of $M(r,\mathbf{v}_r)$ be denoted by $G_r(\mathbf{v}_r)$. From the above definition of $M(r,\mathbf{v}_r)$ and eq. 2.10, it follows that [CON3]

$$G_r(\mathbf{v}_r) =$$

$$\sum_{\mathbf{n}^{(r)} \in S^{(r)}} \prod_{i=1}^{P} \{[\prod_{a=1}^{n_i^{(r)}} (a+v_{ir})] \prod_{s=1}^{r} w_{is}^{n_{is}}/n_{is}!\}\{\prod_{i=P+1}^{N} \prod_{s=1}^{r} w_{is}^{n_{is}}/n_{is}!\},$$

where $n_i^{(r)} = \sum_{s=1}^{r} n_{is}$, nodes 1,...,P are assumed, without loss of generality, to be of the FCFS, LCFSPR or PS types, and nodes (P+1),...,N are assumed to be of the IS type. Since there is never any contention among customers at an IS queue, we note that it is possible to aggregate all queues of the IS type into a single IS queue by suitable specification of the routing-parameters and service-time requirements. Hence, we may assume, for the sake of simplicity that P = N+1.

In [CON3], it has been proved by a simple algebraic method that

$$G_r(\mathbf{v}_r) = \sum_{\mathbf{c} \in L_r} g_r(\mathbf{v}_r,\mathbf{c})G_{r-1}(\mathbf{v}_r+\mathbf{c}), \tag{5.8}$$

where $\mathbf{c} = (c_1,...,c_N)$, $L_r = \{\mathbf{c} \mid c_i \in \{0,1,2,...\}$ for $1 \leq i \leq N$, $\sum_{i=1}^{N} c_i = K_r\}$,

$$g_r(\mathbf{v}_r,\mathbf{c}) = \prod_{i=1}^{N} h_{ir}(\mathbf{v}_r,\mathbf{c})$$

and

$$h_{ir}(\mathbf{v}_r,\mathbf{c}) = \begin{cases} \binom{c_i+v_{ir}}{v_{ir}} w_{ir}^{c_i}, & \text{if node i is FCFS, LCFSPR or PS,} \\ w_{ir}^{c_i}/c_i!, & \text{if node i is IS.} \end{cases}$$

In [DOR1], it has been shown that eq. 5.8 may also be derived using multi-dimensional z-transforms.

The normalization constant of the network B(N,**K**) of interest is, using the notation that we have adopted, equal to $G_R(\mathbf{0})$ since if $\mathbf{v}_R$ = (0,...,0), then $\beta_i(n_i,0) = 1$. This corresponds to the case of constant service-rate functions, which is the situation that we are considering initially. The quantity $G_R(\mathbf{0})$ may be computed recursively, using eq. 5.8, starting with the initial conditions $G_0(\mathbf{v}_0) = 1$. This procedure is summarized below. In the following, the set V_r is defined to be

$$V_r = \begin{cases} \{\mathbf{v}_r \mid v_{ir} \in \{0,1,2,...\} \text{ for } 1 \leq i \leq N, \sum_{i=1}^{N} v_{ir} = \sum_{s=r+1}^{R} K_s\}, & \text{if } 0 \leq r \leq R-1, \\ \{\mathbf{0}\}, \text{ if } r = R. \end{cases}$$

Computation of $G_R(0)$ using eq. 5.8:

Initialization: For all $\mathbf{v}_0 \in V_0$:
$G_0(\mathbf{v}_0) = 1.$

Main Recursion: For $r = 1,...,R$:
For all $\mathbf{v}_r \in V_r$:
Compute $G_r(\mathbf{v}_r)$ using eq. 5.8.

The final result of the above summarized algorithm is $G_R(\mathbf{0})$ since $V_R = \{\mathbf{0}\}$. An efficient method of implementing this recursive algorithm has been formulated in [CON5].

Although the above algorithm may be used to compute the quantity $G_R(\mathbf{0})$ in an efficient manner, it is not clear how one may obtain the mean performance measures in terms of $G_R(\mathbf{0})$ and the normalization constants that have been obtained in the course of computing it. However, a simple method to bypass this difficulty has been devised in [CON3]. The main idea is to suppose that in each routing chain we have a single customer. This assumption involves *no* loss in generality since a routing chain r consisting of K_r identical customers is entirely equivalent physically to K_r single-customer routing chains that have identical routing-parameters and mean service-time requirements. Hence, in the remaining part of this section, we adopt the assumption that $K_r = 1$ for $1 \le r \le R$. In this situation, $L_r = \{\mathbf{1}_i \mid 1 \le i \le N\}$, where $\mathbf{1}_i$ is a unit vector pointing in the direction i, and

$$V_r = \{\mathbf{v}_r \mid v_{ir} \in \{0,1,2,...\} \text{ for } 1 \le i \le N, \sum_{i=1}^{N} v_{ir} = R-r\}.$$

Furthermore, assuming that $K_r = 1$, eq. 5.8 simplifies to

$$G_r(\mathbf{v}_r) = \sum_{i=1}^{N} (1+v_{ir}\delta_i)G_{r-1}(\mathbf{v}_r+\mathbf{1}_i), \tag{5.9}$$

where

$$\delta_i = \begin{cases} 1, & \text{if node i is FCFS, LCFSPR or PS,} \\ 0, & \text{if node i is IS.} \end{cases}$$

If we now compute $G_R(\mathbf{0})$, following the algorithm summarized above using eq. 5.9, instead of eq. 5.8, then we may obtain readily the mean performance measures for the single customer belonging to the routing chain labelled R as follows [CON3]:

$$T_{iR} = \begin{cases} e_{iR} \sum_{j=1}^{N} G_{R-1}(\mathbf{1}_j)/(G_R(\mathbf{0})(N+K-1)), \text{ if } N = P, \\ \qquad \text{(no IS queues in the network)} \\ e_{iR}G_{R-1}(\mathbf{1}_N)/G_R(\mathbf{0}), \text{ if } N > P, \\ \qquad \text{(there is an IS queue in the network)} \end{cases}$$

where K is the total population of customers in the network,

$U_{iR} = t_{iR}T_{iR}$,

$Q_{iR} = w_{iR}G_{R-1}(\mathbf{1}_i)/G_R(\mathbf{0})$

and

$W_{iR} = Q_{iR}/T_{iR}.$

These are the mean performance measures for chain R assuming that $K_R = 1$. If the total number of single-customer routing chains, identical to chain R, is actually $K_R{}^*$, then the performance measures for the aggregated routing chain (to be labelled R*) containing $K_R{}^*$ identical customers may be obtained immediately as follows:

$T_{iR}{}^* = K_R{}^*T_{iR},$

$U_{iR}{}^* = K_R{}^*U_{iR},$

$Q_{iR}{}^* = K_R{}^*Q_{iR}$

and

$W_{iR}{}^* = W_{iR}.$

In order to obtain the mean performance measures for other particular routing chains of interest, it suffices to simply recompute $G_R(\mathbf{0})$, using the same algorithm, but with a new enumeration of the routing chains that assigns the label R to another particular routing chain of interest. This procedure of reenumeration and recomputation of $G_R(\mathbf{0})$ may be repeated until the mean performance measures of all of the distinguishable routing chains in $B(N,\mathbf{K})$ have been obtained. An efficient method to implement the algorithm and carry out this repetitive process is described in detail in [CON3]. The algorithm summarized above has been termed the Recursion by Chain Algorithm since the underlying equation (eq. 5.8) is a recursion over r, the number of closed routing chains.

We now consider how the algorithm summarized above may be modified in order to accommodate state-dependent service-rate functions of the form $\mu_i(n_i)$. In this situation, the queueing network $M(r,\mathbf{v}_r)$ is one that consists of the first r routing chains of $B(N,\mathbf{K})$, but in which there are *special* state-dependent service-rate functions of the form

$$\beta_i(x,v_{ir}) = x\mu_i(x+v_{ir})/(x+v_{ir}).$$

With this change, eq. 5.9 becomes [CON3]

$$G_r(\mathbf{v}_r) = \sum_{i=1}^{N} (1+v_{ir}\delta_i)G_{r-1}(\mathbf{v}_r+\mathbf{1}_i)/\mu_i(1+v_{ir})$$

and we have [CON3]

$$Q_{iR} = w_{iR}G_{R-1}(\mathbf{1}_i)/(\mu_i(1)G_R(\mathbf{0})).$$

In the case for which N > P (at least one IS node in the network), we have [CON3]

$$T_{iR} = e_{iR}G_{R-1}(\mathbf{1}_N)/G_R(\mathbf{0})$$

and

$$W_{iR} = Q_{iR}/T_{iR}.$$

In the case of constant service-rate functions, the space requirement of RECAL is [CON3]

$$\binom{K+N-1}{N-1}+\binom{R+N-1}{N-1} \tag{5.10}$$

and the time requirement (number of additions and multiplications) is [CON3]

$$(4N-1)[\binom{K+N-1}{N}+\binom{R+N-1}{N+1}+\binom{R+N-1}{N}-1]+R(N+8).$$

In the above, R is the number of *distinct* closed routing chains in the network and K is the *total* population of the network. In the case of state-dependent service-rate functions, the space requirement is the same as eq. 5.10 and the time requirement is [CON3]

$$(5N-4)[\binom{K+N-1}{N}+\binom{R+N-1}{N+1}+\binom{R+N-1}{N}-1]+8R.$$

If we suppose that in the *original* network $B(N,\mathbf{K})$ $K_r = \kappa$, for $1 \leq r \leq R$, and consider N and κ as fixed quantities, then, as $R \rightarrow \infty$, the storage requirement of RECAL is of the order of R^{N-1} and the time requirement is of the order of R^{N+1} [CON3].

The Recursion by Chain Algorithm may also be used to analyze mixed queueing network in which there are constant service rates or limited queue-dependent service-rate functions. This is done by transforming the mixed network into an equivalent closed network and analyzing the closed network using RECAL in a way analogous to analyzing mixed networks using the Convolution Algorithm, as described in Section 5.1. If there is class-switching in the network, then we may also immediately obtain the mean performance measures on a per-class basis using eqs. 5.6. Finally, we mention that, like in the Convolution algorithm, it is quite possible that the floating-point range of a machine may be exceeded in the course of computing $G_R(\mathbf{0})$. To reduce the possibility of this occurrence, a simple *dynamic scaling* procedure has been devised in [CON3] that can be incorporated easily within the basic algorithm. Practical experience has shown that this dynamic scaling procedure is very effective [MCK3].

5.4 MVAC - Mean Value Analysis by Chain

In Table 1.1, we provided a general classification of the main computational algorithms that have been developed to date. As can be seen from this table, the newly developed *Mean Value Analysis by Chain* (MVAC) algorithm has essentially the same storage and computational requirements as RECAL. In contrast to RECAL, however, the underlying recursions used in MVAC are in terms of mean performance measures and, in certain situations, marginal queue-length distributions. If we regard RECAL as being the dual of the Convolution Algorithm, then we may regard MVAC as the dual of the MVA algorithm. Consequently, MVAC has the same advantages relative to RECAL as the MVA algorithm has relative to the Convolution Algorithm, namely, its inherent numerical stability and pedagogical appeal.

There are two closely related variations of the MVAC algorithm that have been developed. The first variation, which has been presented in [CON6,SOU1], may be termed *Sequential MVAC* since, in a single pass of the basic recursion employed, we only obtain the mean performance measures for a particular routing chain R. The measures for other routing chains of interest are obtained sequentially by essentially repeating the basic algorithm but with a new enumeration of the routing chains, as is done in RECAL. In the other variation, termed *Parallel MVAC*, the mean performance measures of all of the routing chains in $B(N,\mathbf{K})$ are obtained in a single pass. The parallel version was developed in [CON2,CON9]. The sequential version and the parallel version have essentially the same storage and computational requirements. In the following, we shall first consider the sequential version of MVAC.

5.4.1 Sequential MVAC

There are two formulations of the so-called Sequential MVAC algorithm. The first formulation is applicable to multiple-chain BCMP queueing networks with constant service-rate functions and the second is designed to accommodate state-dependent service-rate functions of the form $\mu_i(n_i)$. The latter version may, of course, also be used in the case of constant service-rates. In both versions, we consider the queueing network $B(N,\mathbf{K})$ and associate a separate closed routing chain with each customer in the network. Hence, in the following we assume, without loss of generality, that $K_r = 1$ for $1 \leq r \leq R$. If in the original queueing network $B(N,\mathbf{K})$ under consideration there are multiple customers in the routing chains, then, as has been explained in Section 5.3, we may break down each of these multiple-customer routing chains into single-customer ones. Since this transformation has no physical effect, the mean performance measures remain unchanged.

Consider again the queueing network $M(r,\mathbf{v})$, where $\mathbf{v} = (v_1,...,v_N)$, as defined previously in Section 5.3, and suppose, initially, that we have constant service-rate functions. The network $M(r,\mathbf{v})$ is one that consists of the first r routing chains of $B(N,\mathbf{K})$ and in which there are special state-dependent service-rate functions of the form

$$\beta_i(x,v_i) = x/(x+v_i).$$

An alternate and equivalent interpretation of $M(r,\mathbf{v})$ is as a queueing network with $\beta_i(x,v_i) = 1$, but with v_i so-called *single-customer self-looping* (SCSL) chains at node i, where $1 \leq i \leq N$. A SCSL chain is defined to be a single customer that loops around continuously at node i. If we assume further, without loss of generality, that all queues in $M(r,\mathbf{v})$ of the FCFS and LCFSPR types are converted to PS, then the actual service-time requirements of the customers in the SCSL chains are redundant since, when the customer of a SCSL chain finishes service at node i, it returns *instantaneously* back to node i for another independent round of service. Furthermore, the conversion of FCFS and LCFSPR queues, to queues of the PS type, has no effect on the mean performance measures since, as can be seen from eq. 2.10, all of these queueing disciplines exhibit identical marginal state-distributions. In the remaining developments to be made in this section, we choose to adopt the interpretation of $M(r,\mathbf{v})$ in terms of SCSL chains since this facilitates the probabilistic development of the underlying recursive equations of MVAC.

Consider the queueing network $M(r,\mathbf{v})$, where $\mathbf{v} = (v_1,...,v_N)$ and v_i is the number of SCSL chains at node i. In terms of this notation, the queueing network $B(N,\mathbf{K})$ of interest is equivalent to $M(R,\mathbf{0})$ since when $\mathbf{v} = \mathbf{0}$, there are no SCSL chains at any of the nodes of the network. Let $Q_{is}(r,\mathbf{v})$, $1 \le s \le r$, be the mean number of chain s customers at node i in the network $M(r,\mathbf{v})$. In [CON6,SOU1], it has been established that

$$Q_{is}(r,\mathbf{v}) = \sum_{j=1}^{N} Q_{jr}(r,\mathbf{v})Q_{is}(r-1,\mathbf{v}+\mathbf{1}_j),$$

where $1 \le s \le (r-1)$. In the above, the quantity $Q_{jr}(r,\mathbf{v})$ may be interpreted as the probability that the single customer of chain r in $M(r,\mathbf{v})$ is at node j and $Q_{is}(r-1,\mathbf{v}+\mathbf{1}_j)$ may be interpreted as the conditional probability that the single customer of chain s, $1 \le s \le r-1$, is at node i in $M(r,\mathbf{v})$, given that the single customer of chain r is at node j [CON6]. Hence,

$$Q_i(r,\mathbf{v}) = Q_{ir}(r,\mathbf{v}) + \sum_{j=1}^{N} Q_{jr}(r,\mathbf{v})Q_i(r-1,\mathbf{v}+\mathbf{1}_j), \tag{5.11}$$

where

$$Q_i(r,\mathbf{v}) = \sum_{s=1}^{r} Q_{is}(r,\mathbf{v}) \quad \text{and} \quad Q_i(r-1,\mathbf{v}+\mathbf{1}_j) = \sum_{s=1}^{r-1} Q_{is}(r-1,\mathbf{v}+\mathbf{1}_j).$$

Now let $W_{jr}(r,\mathbf{v})$ be the mean waiting-time (including service) for the single customer of chain r at node j in the network $M(r,\mathbf{v})$. Then, using the waiting-time equation for the case of constant service-rate functions, as given in eqs. 5.7, and recognizing that there are v_j SCSL chains at node j, it follows immediately that, in terms of the notation adopted in this section,

$$W_{jr}(r,\mathbf{v}) = \begin{cases} t_{jr}(1+Q_j(r-1,\mathbf{v})+v_j), & \text{if node j is FCFS, LCFSPR or PS,} \\ t_{jr}, & \text{if node j is IS.} \end{cases}$$

Hence, by Little's result [LIT1],

$$Q_{jr}(r,\mathbf{v}) = \begin{cases} w_{jr}T(r,\mathbf{v})(1+Q_j(r-1,\mathbf{v})+v_j), & \text{if node j is FCFS, LCFSPR or PS,} \\ w_{jr}T(r,\mathbf{v}), & \text{if node j is IS,} \end{cases} \quad (5.12)$$

where $T(r,\mathbf{v}) = e_{jr}^{-1}T_{jr}(r,\mathbf{v})$ and $T_{jr}(r,\mathbf{v})$ is the throughput of the single customer of chain r at node j in the network $M(r,\mathbf{v})$. Furthermore, by definition,

$$\sum_{j=1}^{N} Q_{jr}(r,\mathbf{v}) = 1$$

and so we have

$$T(r,\mathbf{v}) = 1 \;/\; [\sum_{j=1}^{P} w_{jr}(Q_j(r-1,\mathbf{v})+v_j) + \sum_{j=1}^{N} w_{jr}], \quad (5.13)$$

where nodes 1,...,P are assumed to be of the PS type (or FCFS, LCFSPR) and nodes P+1,...,N are assumed to be of the IS type.

Equations 5.11, 5.12 and 5.13 constitute a set of recursive equations that may be used to obtain the mean performance measures for chain R in the network $B(N,\mathbf{K})$, assuming constant service-rate functions. The algorithm may be summarized as below. In the following, the set V_r is defined to be

$$V_r = \{\mathbf{v}_r \mid \mathbf{v}_r = (v_{1r},...,v_{Nr}),\ v_{ir} \in \{0,1,2,...\} \text{ for } 1 \le i \le N,\ \sum_{i=1}^{N} v_{ir} = K\text{-}r\},$$

where K (K = R) is the total number of customers in B(N,**K**).

Sequential MVAC Algorithm - Constant Service-Rate Functions:

Initializations: For r = 1,...,K:
For all $\mathbf{v} \in V_r$:
$Q_j(0,\mathbf{v}) = 0$, for $1 \le j \le N$.

Main Recursion: For r = 1,...,K:
For all $\mathbf{v} \in \{V_r, V_{r+1}, ..., V_K\}$:
Find $T(r,\mathbf{v})$ using eq. 5.13.
Find $Q_{jr}(r,\mathbf{v})$, for $1 \le j \le N$, using eq. 5.12.
Find $Q_i(r,\mathbf{v})$, for $1 \le i \le N$, using eq. 5.11.

Since $V_K = \{\mathbf{0}\}$, the final results of the algorithm summarized above are $Q_{jR}(R,\mathbf{0})$, for $1 \le j \le N$, $T(R,\mathbf{0})$ and $Q_i(R,\mathbf{0})$, for $1 \le i \le N$. We also have $T_{jR}(R,\mathbf{0})$ since

$$T_{jR}(R,\mathbf{0}) = e_{jR}T(R,\mathbf{0}).$$

These are the mean performance measures for chain R, assuming a single customer in each routing chain. If, in actual fact, there are a total of K_R^* routing chains identical to chain R, then the performance measures for the aggregated routing chain (labelled R^*) may be obtained immediately as follows:

$$T_{iR}^* = K_R^* T_{iR}(R,\mathbf{0}),$$

$U_{iR}^* = t_{iR}T_{iR}^*,$

$Q_{iR}^* = K_R^*Q_{iR}(R,\mathbf{0})$

and

$$W_{iR}^* = W_{iR}(R,\mathbf{0}). \tag{5.14}$$

In order to obtain the mean performance measures for other routing chains of interest, it suffices simply to repeat the above algorithm but with a new enumeration of the routing chains that assigns the label R to other particular routing chains of interest, as was done in RECAL. This sequential repetition of the algorithm may be carried out until the mean performance measures for all of the distinct routing chains in $B(N,\mathbf{K})$ have been obtained. An efficient method for implementing this sequential process is described in [CON6,SOU1].

We now consider the second formulation of the Sequential MVAC algorithm that may be applied to queueing networks with state-dependent service-rate functions of the form $\mu_i(n_i)$. As before, we assume that $K_r = 1$ for $1 \le r \le R$ so that we have $K = R$. Again, let $M(r,\mathbf{v})$ denote a queueing network that consists of the first r chains of $B(N,\mathbf{K})$ and in which there are v_i SCSL chains located at node i, where $1 \le i \le N$. We now assume that at node i in $M(r,\mathbf{v})$ we have the state-dependent service-rate function $\mu_i(n_i)$. In [CON6,CON9], it has been established that

$$P_j(r,n,\mathbf{v}) = Q_{jr}(r,\mathbf{v})P_j(r-1,n-1,\mathbf{v}+\mathbf{1}_j) + \sum_{\substack{i=1 \\ i \ne j}}^{N} Q_{ir}(r,\mathbf{v})P_j(r-1,n,\mathbf{v}+\mathbf{1}_i), \tag{5.15}$$

where $P_j(r-1,n,\mathbf{v}+\mathbf{1}_j) = 0$, if $n > (r-1)$, $P_j(0,0,\mathbf{v}) = 1$ and $P_j(r,n,\mathbf{v})$ is the probability that there are a total of n customers at node j in the network $M(r,\mathbf{v})$. In the above, $P_j(r-1,n,\mathbf{v}+\mathbf{1}_i)$ may be interpreted as the conditional probability that the total number of customers at node i in M(r,**v**) is n, given that the single customer of chain r is at node i. From [CON6], we also have that

$$Q_{jr}(r,\mathbf{v}) = e_{jr}\tau_r(\mathbf{v},j)^{-1} \Big/ \sum_{i=1}^{N} e_{ir}\tau_r(\mathbf{v},i)^{-1}, \tag{5.16}$$

where

$$\tau_r(\mathbf{v},i) = \begin{cases} t_{ir}^{-1} \displaystyle\sum_{n=0}^{r-1} P_i(r-1,n,\mathbf{v}+\mathbf{1}_i)\mu_i(n+v_i+1)/(n+v_i+1), & \text{if node i is PS (or FCFS, LCFSPR),} \\ t_{ir}^{-1}, \text{ if node i is IS,} & \end{cases}$$

and

$$T_{jr}(r,\mathbf{v}) = \tau_r(\mathbf{v},j)Q_{jr}(r,\mathbf{v}). \tag{5.17}$$

In the above, $\tau_r(\mathbf{v},i)^{-1}$ may be interpreted as the mean waiting-time of the single customer of chain r at node i in M(r,**v**).

Equations 5.15, 5.16 and 5.17 constitute a set of recursive equations that may be used to obtain the mean performance measures for chain R in B(N,**K**) when we have state-dependent service-rate functions. The resulting algorithm may be summarized as follows:

Sequential MVAC Algorithm - State-Dependent Service-Rate Functions:

Initializations: For all $\mathbf{v} \in V_R$:

$P_i(0,n,\mathbf{v}) = 0$, for $1 \le i \le N$, $n > 0$.

$P_i(0,0,\mathbf{v}) = 1$.

Main Recursion: For r = 1,...,R:

For all $\mathbf{v} \in V_r$:
Find $Q_{jr}(r,\mathbf{v})$, for $1 \leq j \leq N$, using eq. 5.16.
Find $P_j(r,n,\mathbf{v})$, for $1 \leq j \leq N$ and $0 \leq n \leq r$, using eq. 5.15.
Find $T_{jR}(R,\mathbf{0})$, for $1 \leq j \leq N$, using eq. 5.17.

The final results obtained by the above algorithm are $Q_{jR}(R,\mathbf{0})$ and $T_{jR}(R,\mathbf{0})$, for $1 \leq j \leq N$, and $P_j(R,n,\mathbf{0})$, for $1 \leq j \leq N$ and $1 \leq n \leq K$. Hence, we have the mean performance measures for chain R in $M(R,\mathbf{0})$ or $B(N,\mathbf{K})$. The mean performance measures for the other routing chains in $B(N,\mathbf{K})$ may be obtained using the same sequential procedure described above for the formulation of the Sequential MVAC algorithm that assumes constant service-rate functions.

We now summarize the storage requirements and computational costs of the Sequential MVAC algorithm [CON6,SOU1]. If we suppose that in the original network $K_r = \kappa$, for $1 \leq r \leq R$, and consider N and κ as fixed, then in the case of constant service-rate functions the storage requirement of MVAC, as $R \to \infty$, is of the order of R^N and the time requirement is of the order of a function $\vartheta(R)$, where $R^{N+1} \leq \vartheta(R) \leq R^{N+2}$. In the case of state-dependent service-rate functions, the maximum time requirement of MVAC is of the order of R^{N+2} and the storage requirement is of the order of R^N. More detailed formulae for the space and time requirements of MVAC are given in [CON6,SOU1].

5.4.2 Parallel MVAC

The Sequential MVAC algorithm requires that a number of passes be made through the basic recursion in order to obtain the mean performance measures for all of the routing chains in $B(N,\mathbf{K})$. We now describe an algorithm that obtains all of the mean performance measures in parallel, in a single pass. This so-called Parallel MVAC algorithm is closely related to the version of the Sequential MVAC algorithm that applies to networks with state-dependent service-rate functions.

Consider the queueing network $M(r,\mathbf{v})$ and suppose, as before, that there is a single customer in each routing chain. Also assume that there are state-dependent service-rate functions of the form $\mu_i(n_i)$. From eq. 5.15, we have

$$P_j(r,n,\mathbf{v}) = Q_{jr}(r,\mathbf{v})P_j(r-1,n-1,\mathbf{v}+\mathbf{1}_j) + \sum_{\substack{i=1 \\ i \neq j}}^{N} Q_{ir}(r,\mathbf{v})P_j(r-1,n,\mathbf{v}+\mathbf{1}_i) \qquad (5.18)$$

and, from eq. 5.16, we have

$$Q_{jr}(r,\mathbf{v}) = e_{jr}\tau_r(\mathbf{v},j)^{-1} \Big/ \sum_{i=1}^{N} e_{ir}\tau_r(\mathbf{v},i)^{-1}, \qquad (5.19)$$

where

$$\tau_r(\mathbf{v},i) = \begin{cases} t_{ir}^{-1} \sum_{n=0}^{r-1} P_i(r-1,n,\mathbf{v}+\mathbf{1}_i)\mu_i(n+v_i+1)/(n+v_i+1), \\ \qquad\qquad \text{if node } i \text{ is PS (or FCFS, LCFSPR),} \\ t_{ir}^{-1}, \text{ if node } i \text{ is IS.} \end{cases}$$

In [CON9], it has been proved that

$$T_{is}(r,\mathbf{v}) = \sum_{j=1}^{N} Q_{jr}(r,\mathbf{v})[T_{is}(r-1,\mathbf{v}+\mathbf{1}_j) + \delta_{ijsr}\tau_r(\mathbf{v},j)], \qquad (5.20)$$

where $1 \leq i \leq N$, $1 \leq s \leq r$, $T_{is}(r-1,\mathbf{v}+\mathbf{1}_j) = 0$ if $s > (r-1)$, and

$$\delta_{ijsr} = \begin{cases} 1, & \text{if } i = j \text{ and } s = r, \\ 0, & \text{otherwise} \end{cases}$$

In eq. 5.20, $Q_{jr}(r,\mathbf{v})$ is the probability that the single customer of chain r in $M(r,\mathbf{v})$ is at node j, and $T_{is}(r-1,\mathbf{v}+\mathbf{1}_j)$ may be interpreted as the conditional throughput of the single customer of chain s at node i in $M(r,\mathbf{v})$, given that the single customer of chain r is at node j. The quantity $\tau_r(\mathbf{v},j)$ may be interpreted as the conditional transition rate of the single customer of chain r out of node j, given that it is at node j. It has also been proved in [CON9] that

$$Q_{is}(r,\mathbf{v}) = \sum_{j=1}^{N} Q_{jr}(r,\mathbf{v})[Q_{is}(r-1,\mathbf{v}+\mathbf{1}_j) + \delta_{ijsr}], \tag{5.21}$$

where $1 \le s \le r$ and $Q_{is}(r-1,\mathbf{v}+\mathbf{1}_j) = 0$ if $s > (r-1)$. In eq. 5.21, the quantity $Q_{is}(r-1,\mathbf{v}+\mathbf{1}_j)$ may be interpreted as the conditional mean queue-length of chain s at node i in $M(r,\mathbf{v})$, given that the single customer of chain r is at node j.

The above recursive equations may be used to obtain all of the mean performance measures for $B(N,\mathbf{K})$ in a single pass. This Parallel MVAC algorithm may be summarized as follows:

Parallel MVAC Algorithm:

Initializations:

For all $\mathbf{v} \in V_R$:
$P_i(0,n,\mathbf{v}) = 0$, for $1 \le i \le N$, $n > 0$.
$P_i(0,0,\mathbf{v}) = 1$.
$Q_{is}(0,\mathbf{v}) = 0$, for $1 \le i \le N$, $1 \le s \le R$.
$T_{is}(0,\mathbf{v}) = 0$, for $1 \le i \le N$, $1 \le s \le R$.

Main Recursion:

For $r = 1,\ldots,R$:
For all $\mathbf{v} \in V_r$:
Find $Q_{jr}(r,\mathbf{v})$, for $1 \le j \le N$,

using eq. 5.19.
Find $P_j(r,n,\mathbf{v})$, for $1 \le j \le N$,
and $0 \le n \le r$, using eq. 5.18.
Find $T_{is}(r,\mathbf{v})$, for $1 \le j \le N$ and $1 \le s \le r$, using eq. 5.20.
Find $Q_{is}(r,\mathbf{v})$, for $1 \le j \le N$ and $1 \le s \le r$, using eq. 5.21.

The final results obtained by the Parallel MVAC algorithm are $T_{is}(R,\mathbf{0})$ and $Q_{is}(R,\mathbf{0})$, for $1 \le j \le N$ and $1 \le s \le R$, and $P_j(R,n,\mathbf{0})$ for $1 \le j \le N$ and $0 \le n \le R$. From Little's result [LIT1], we also have

$$W_{is}(R,\mathbf{0}) = Q_{is}(R,\mathbf{0})/T_{is}(R,\mathbf{0}).$$

These are all of the mean performance measures of interest since the queueing network $M(R,\mathbf{0})$ is, by definition, identical to $B(N,\mathbf{K})$.

We now consider the storage and computational requirements of the Parallel MVAC algorithm. The total storage requirement is [CON9]

$$(N(K+1)+2NR+2N)\binom{K+N-1}{N-1},$$

where K is the total population of the network and R is the number of *distinct* routing chains. The time requirement (number of multiplications, divisions and additions) is [CON9]

$$(N(2N+3)(K+1)+2RN(2N-1)-N^2(2N+3)(K-1)/(N+1)+2)\binom{K+N-1}{N}.$$

If we suppose that in the original network $K_r = \kappa$, for $1 \le r \le R$, and consider N and κ as fixed, then, as $R \to \infty$, the storage requirement of the Parallel MVAC algorithm is of the order of R^N and the time requirement is of the order of R^{N+1}.

5.5 DAC - Distribution Analysis by Chain

A new computational algorithm that has been developed very recently by de Souza e Silva and Lavenberg [SOU2] is the so-called *Distribution Analysis by Chain* (DAC) algorithm. In contrast to the other exact algorithms that we have seen in this chapter, the recursion employed in DAC is in terms of the *joint* marginal queue-length distribution of networks of the type defined by $B(N,\mathbf{K}_r)$, or equivalently, $M(r,\mathbf{0})$, where $1 \leq r \leq R$ and $\mathbf{K}_r = (K_1,...,K_r)$. Although the DAC algorithm was developed especially for the purpose of computing joint queue-length distributions, such as those that are required in the evaluation of system availability models of the type described in Subsection 2.3.8, the algorithm may also be used to obtain the mean performance measures of BCMP queueing networks. Hence, the DAC algorithm may be regarded as another general purpose exact computational algorithm.

The DAC algorithm may be classified as a MVA type of algorithm since it does not involve the computation of normalization constants. The storage requirements and computational costs of DAC are closely related to those of RECAL and MVAC. As a result, the DAC algorithm is useful when one is interested in obtaining the mean performance measures of networks in which there are many routing chains and few nodes or when it is required that the joint marginal queue-length distribution be obtained. In [SOU2], it has been established theoretically that the DAC algorithm is actually more efficient than either RECAL or MVAC.

Consider the queueing network $B(N,\mathbf{K})$ and assume that we have state-dependent service-rate functions of the form $\mu_i(n_i)$ at the nodes. As was done in the developments of RECAL and MVAC, we assume, in the following, that there is a closed routing chain associated with each individual customer in the queueing network, even though two or more customers in the network may, in fact, be indistinguishable as far as their routing and service-requirement parameters are concerned. Hence, in the following, we assume that K_r

$= 1$, for $1 \le r \le R$, so that $R = K$, where $K = K_1+\ldots+K_R$. Now let $P^r(\mathbf{n})$ be the probability that $\mathbf{n} = (n_1,\ldots,n_N)$, where $n_i = n_{i1}+\ldots+n_{ir}$ and n_{is} is the number of customers of chain s at node i in the network $B(N,\mathbf{K}_r)$. The joint marginal queue-length distribution for this network is then

$$\{P^r(\mathbf{n}) \mid \mathbf{n} = (n_1,\ldots,n_N), \mathbf{n} \in L_r'\},$$

where

$$L_r' = \{ \mathbf{n} \mid \mathbf{n} = (n_1,\ldots,n_N), 0 \le n_i \le r, \sum_{i=1}^{N} n_i = r\}.$$

In [SOU2], it has been proved that

$$P^r(\mathbf{n}) = T(r) \sum_{j=1}^{N} w_{jr} n_j P^{r-1}(\mathbf{n}-\mathbf{1}_j)/\mu_j(n_j), \tag{5.22}$$

where

$$T(r) = e_{jr}^{-1} T_{jr}(r),$$

$$T(r) = 1 / [\sum_{j=1}^{N} w_{jr} \sum_{x=1}^{r} x \; P_j^{r-1}(x-1)/\mu_j(x)] \tag{5.23}$$

and

$$P_j^{r-1}(x) = \sum_{\substack{\mathbf{n} \in L_{r-1}' \\ n_j = x}} P^{r-1}(\mathbf{n}). \tag{5.24}$$

The quantity $P^{r-1}(\mathbf{n})$ is the probability that the joint queue-length distribution is $\mathbf{n}$ in $B(N,\mathbf{K}_{r-1})$ and $T_{jr}(r)$ is the throughput of chain r at node j in $B(N,\mathbf{K}_r)$. From [SOU2], we also have that

$$Q_{jr}(r) = T(r)w_{jr}\sum_{x=1}^{r} xP_j^{r-1}(x-1)/\mu_j(x), \tag{5.25}$$

where $Q_{jr}(r)$ is the mean queue-length of chain r at node j in $B(N,\mathbf{K}_r)$.

Equations 5.22, 5.23, 5.24 and 5.25 constitute a set of recursive equations that may be used to obtain the joint queue-length distribution for the network $B(N,\mathbf{K}_R)$, assuming state-dependent service-rate functions. The set of equations also yields the mean performance measures $T_{jR}(R)$ and $Q_{jR}(R)$ for chain R. The initial conditions for the recursion are $P_j^1(1)$ and $P_j^1(0)$, for $1 \le j \le N$. From [SOU2], we have that

$$P_j^1(1) = w_{j1}/\sum_{i=1}^{N} w_{i1}. \tag{5.26}$$

We now summarize the algorithm based on the above set of recursive equations.

DAC - Distribution Analysis by Chain:

Initializations:

For $1 \le j \le N$:
- Find $P_j^1(1)$ using eq. 5.26.
- $P_j^1(0) = 1 - P_j^1(1)$.

Main Recursion:

For $r = 2,3,...,K$:
- Find $T(r)$ using eq. 5.23.
- For all $\mathbf{n} \in L_r'$:
  - Find $P^r(\mathbf{n})$ using eq. 5.22.
- For $1 \le j \le N$ and $0 \le x \le r$:
  - Find $P_j^r(x)$ using eq. 5.24.

For $1 \le j \le N$:
- $T_{jR}(R) = e_{jR}T(R)$.
- Find $Q_{jR}(R)$ using eq. 5.25.
- $W_{jR}(R) = Q_{jR}(R)/T_{jR}(R)$.

The above summarized algorithm yields the joint distribution $P^R(\mathbf{n})$ for the network $B(N,\mathbf{K}_R)$ but only the mean performance measures $T_{jR}(R)$, $Q_{jR}(R)$ and $W_{jR}(r)$ for chain R.

To obtain the mean performance measures for other routing chains of interest in $B(N,\mathbf{K}_R)$, it suffices to simply repeat the algorithm but with a new enumeration of the routing chains that assigns the label R to another particular chain of interest. It is, however, possible to obtain these other mean performance measures in a more straightforward way, assuming that one has already obtained the joint distribution [SOU2]. This may be done as follows.

Consider the queueing network $B(N,\mathbf{K}_R)$ and suppose that we reenumerate the routing chains so that chain s, $1 \leq s \leq R-1$, is now assigned the label R. Then, from eq. 5.22, we have

$$P^R(\mathbf{n}) = T(s)\sum_{j=1}^{N} w_{js}n_jP^{R-1,s}(\mathbf{n}-\mathbf{1}_j)/\mu_j(n_j), \tag{5.27}$$

where $P^{R-1,s}(\mathbf{n}-\mathbf{1}_j)$ is used to denote the joint queue-length distribution of a network which is identical to $B(N,\mathbf{K}_R)$, but with chain s removed.

Now, by definition, let $Z^{R-1,s}(\mathbf{n}-\mathbf{1}_j) = T(s)P^{R-1,s}(\mathbf{n}-\mathbf{1}_j)$. It immediately follows from eq. 5.27 that

$$\sum_{j=1}^{N} w_{js}n_jZ^{R-1,s}(\mathbf{n}-\mathbf{1}_j)/\mu_j(n_j) = P^R(\mathbf{n}) \tag{5.28}$$

Equation 5.28 may be used to find $Z^{R-1,s}(\mathbf{n}-\mathbf{1}_j)$ for all $\mathbf{n} \in L_R'$ if we determine $Z^{R-1,s}(\mathbf{n}-\mathbf{1}_j)$ in the lexicographic order $\mathbf{n}$ = (K,0,...,0), (K-1,1,0,...,0), (K-2,2,0,...,0), ..., (K-1,0,1,0,...,0), ..., (0,...,0,K). Having found $Z^{R-1,s}(\mathbf{n}-\mathbf{1}_j)$ for all $\mathbf{n} \in L_R'$, we then have

$$T(s) = \sum_{\mathbf{n} \in L_R'} Z^{R-1,s}(\mathbf{n}-\mathbf{1}_j) \quad \text{and} \quad T_{js}(R) = e_{js}T(s) \tag{5.29}$$

since $Z^{R-1,s}(\mathbf{n}-\mathbf{1}_j) = T(s)P^{R-1,s}(\mathbf{n}-\mathbf{1}_j)$ and, by definition,

$$\sum_{\mathbf{n} \in L_{R-1}'} P^{R-1,s}(\mathbf{n}) = 1.$$

Finally, we may obtain $Q_{js}(R)$ using eqs. 5.25 and 5.24.

We now summarize this method of obtaining all of the mean performance measures, assuming that $P^R(\mathbf{n})$ has been previously computed using the DAC algorithm.

Determination of the Performance Measures for Chains 1,...,R-1:

For s = 1,...,R-1:
For all $\mathbf{n} \in L_R'$ in the lexicographic order specified above:
Find $Z^{R-1,s}(\mathbf{n}-\mathbf{1}_j)$ using $P^R(\mathbf{n})$ and the previously found values of $Z^{R-1,s}(\mathbf{n}-\mathbf{1}_x)$, $x \neq j$.

Find T(s) and then $T_{js}(R)$, for $1 \leq j \leq N$, using eqs. 5.29.
Find $P_j^{R-1,s}(x)$, for $1 \leq j \leq N$, $0 \leq x \leq R-1$, using eq. 5.24.
Find $Q_{js}(R)$, for $1 \leq j \leq N$, using eq. 5.25, where in eq. 5.25 T(r) is replaced by T(s).
$W_{js}(R) = Q_{js}(R)/T_{js}(R)$.

We now summarize the computational costs (multiplications, divisions) of the DAC algorithm. To obtain the entire joint queue-length distribution, the cost is [SOU2]

$$(3N+1)\binom{K+N}{N} + NK(K+2) + K - 3N^2 - 6N - 2.$$

To obtain all the mean performance measures using the DAC algorithm in conjunction with the supplementary algorithm described above, if we suppose that in the original network $K_r = \kappa$, for $1 \le r \le R$, and consider N and κ as fixed, then as $R \to \infty$, the computational costs are at most of the order of R^N [SOU2].

We mention, finally, that the DAC algorithm may also be used to analyze networks with class-switching and mixed queueing networks by following the same approach which has been described in Section 5.1 for the Convolution Algorithm.

5.6 Specialized Exact Algorithms

The algorithms that we have described in Sections 5.1 to 5.5 may be regarded as general purpose ones, in the sense that they may be applied to queueing network problems in which no *a priori* constraints are placed on the relative traffic intensities w_{ir}. When there exists certain structure in a queueing network, however, it is sometimes possible to exploit this to construct specialized algorithms with improved efficiency. To construct such specialized algorithms, the approach is to consider an exact algorithm in its general form and study how it may be simplified, given the particular special structure that is assumed.

There are several specialized algorithms that have been developed to date. These include the *Tree Convolution* algorithm [LAM3], *Tree MVA* [HOY1,TUC1], the *Tree RECAL Algorithm* (TRA) [MCK2,GRE1] and algorithms for the solution of so-called *homogeneous* [BAL2,CON8] and *semi-homogeneous* [CON7] queueing networks. The Tree Convolution algorithm and Tree MVA are particularly useful for the analysis of large queueing network models of computer systems and communication networks in which there are many closed routing chains that exhibit sparsity or locality properties. In such networks, $w_{ir} = 0$ for certain values of i and r. In

Tree Convolution and Tree MVA, this structure is exploited directly to reduce the time and space requirements of the general purpose versions of the respective algorithms. The TRA algorithm is based on the general purpose version of RECAL and also exploits sparsity or locality properties of a network. The algorithms developed in [BAL2,CON7,CON8] are designed to exploit the symmetry (or homogeneity) properties that may sometimes be found in queueing network models of distributed computer systems and local area networks of workstations. The algorithm presented in [BAL2] is based on a general purpose version of the MVA algorithm, while those presented in [CON7,CON8] are based on the general purpose version of RECAL. In the following, we shall only consider the Tree Convolution algorithm in more detail. This will serve to illustrate the process of constructing algorithms with reduced complexity.

Consider the general version of the Convolution Algorithm, as described in Section 5.1. The underlying recursive formula is

$$C_m(\mathbf{k}) = (C_{m-1} \otimes f_m)(\mathbf{k}),$$

where $\mathbf{k} = (k_1,\ldots,k_R)$. More explicitly,

$$C_m(\mathbf{k}) = \sum_{n_R=0}^{k_R} \cdots \sum_{n_2=0}^{k_2} \sum_{n_1=0}^{k_1} f_m(\mathbf{n}) C_{m-1}(\mathbf{k}-\mathbf{n}). \tag{5.30}$$

We may also write

$$C_m(\mathbf{k}) = (f_1 \otimes \ldots \otimes f_m)(\mathbf{k}).$$

Hence, the normalization constant $C_N(\mathbf{K})$ for the network $B(N,\mathbf{K})$ of interest may be written as

$$C_N(\mathbf{K}) = (f_1 \otimes \ldots \otimes f_N)(\mathbf{K}).$$

Now suppose that the set of nodes $\{1,\ldots,N\}$ is partitioned into two disjoint sets of nodes $\sigma_1 = \{1,\ldots,N_1\}$ and $\sigma_2 = \{N_1+1,\ldots,N\}$. Then we may write

$$C_N(\mathbf{K}) = (C_{\sigma_1} \otimes C_{\sigma_2})(\mathbf{K}), \tag{5.31}$$

where

$$C_{\sigma_1}(\mathbf{k}) = (f_1 \otimes \ldots \otimes f_{N_1})(\mathbf{k}) \text{ and } C_{\sigma_2}(\mathbf{k}) = (f_{N_1+1} \otimes \ldots \otimes f_N)(\mathbf{k}).$$

Also, suppose that the routing chains in the network are enumerated in such a way that $\beta_1 = \{1,\ldots,R_1\}$ is the set of routing chains that *only* visit nodes in the set σ_1, $\beta_2 = \{R_1+1,\ldots,R_2\}$ is the set of chains that *only* visit nodes in σ_2 and $\{R_2+1,\ldots,R\}$ is the set of chains that visit nodes in *both* σ_1 and σ_2. That is, if $i \in \sigma_1$, then $e_{ir} = 0$ if $r \in \beta_2$, and if $i \in \sigma_2$, then $e_{ir} = 0$ if $r \in \beta_1$. Now, by definition,

$$C_{\sigma_1}(\mathbf{k}) = \sum_{\mathbf{n}^{(R)} \in S_{\sigma_1}(R,\mathbf{k})} \prod_{i \in \sigma_1} f_i(\mathbf{n}_i^{(R)}),$$

where

$$S_{\sigma_1}(R,\mathbf{k}) = \{\mathbf{n}^{(R)} \mid n_{ir} \geq 0, i \in \sigma_1, \mathbf{k} = (k_1,\ldots,k_R), \sum_{i \in \sigma_1} n_{ir} = k_r, 1 \leq r \leq R\},$$

$$f_i(\mathbf{n}_i^{(R)}) = \begin{cases} n_i^{(R)}! \prod_{r=1}^{R} w_{ir}^{n_{ir}}/n_{ir}!, & \text{if node i is FCFS, PS or LCFSPR,} \\ \prod_{r=1}^{R} w_{ir}^{n_{ir}}/n_{ir}!, & \text{if node i is IS,} \end{cases}$$

$$w_{ir} = e_{ir}t_{ir}, \quad e_{ir} = \sum_{c \in C_{ir}} \alpha_{ic}^{(r)}, \text{ and } t_{ir} = \sum_{c \in C_{ir}} \alpha_{ic}^{(r)} m_{ic}^{(r)}/e_{ir}$$

($\alpha_{ic}^{(r)}$, $m_{ic}^{(r)}$ and C_{ir} have been defined in Subsection 2.3.8). Hence, $C_{\sigma_1}(\mathbf{k}) = 0$ if $k_r > 0$ and $r \in \beta_2$ since $w_{ir} = 0$ when $i \in \sigma_1$ and $r \in \beta_1$. In a similar manner, $C_{\sigma_2}(\mathbf{k}) = 0$ if $k_r > 0$ and $r \in \beta_1$ since $w_{ir} = 0$ when $i \in \sigma_2$ and $r \in \beta_1$. As a result, to obtain $C_N(\mathbf{K})$ using eq. 5.31, the sums need only range over those routing chains r such that $R_2+1 \leq r \leq R$. Hence, eq. 5.31 simplifies to

$$C_N(\mathbf{K}) =$$

$$\sum_{n_{R_2+1}=0}^{K_{R_2+1}} \cdots \sum_{n_R=0}^{K_R} C_{\sigma_1}(K_1,\ldots,K_{R_1},0,\ldots,0,n_{R_2+1},\ldots,n_R)\, C_{\sigma_2}(0,\ldots,0,K_{R_1+1},\ldots,$$
$$K_{R_2},K_{R_2+1}-n_{R_2+1},\ldots,K_R-n_R). \tag{5.32}$$

This is a potentially significant simplification since, to obtain $C_N(\mathbf{K})$ using eq. 5.32, we need only have available the set of normalization constants

$$\{C_{\sigma_1}(K_1,\ldots,K_{R_1},0,\ldots,0,n_{R_2+1},\ldots,n_R), C_{\sigma_2}(0,\ldots,0,K_{R_1+1},\ldots,K_{R_2},n_{R_2+1},\ldots,n_R) \mid$$
$$0 \leq n_s \leq K_s, \text{ where } R_2+1 \leq s \leq R\}.$$

The cardinality of this set is

$$2 \prod_{s=R_2+1}^{R} (K_s+1).$$

In general, the number of normalization constants that would have been required to be available is

$$2 \prod_{r=1}^{R} (K_r+1).$$

Furthermore, the time requirement is reduced significantly since the convolution operation is reduced from R dimensions to $(R-R_2)$ dimensions. These simplifications are the fundamental basis of the Tree Convolution algorithm.

The same observations that we have made above may be applied recursively to the computation of the normalization constants C_{σ_1} and C_{σ_2} appearing in eq. 5.32. To compute these in an efficient manner, we may decompose σ_1 and σ_2 into the pairs of subnetworks σ_{11}, σ_{12} and σ_{21}, σ_{22}, respectively. Then, in the computation of C_{σ_1}, for example, the convolution operation need only range over those chains r that visit nodes both in σ_{11} and σ_{12}. This procedure of network decomposition and computation of normalization constants may be continued until there is a single node in each subnetwork.

Although the above described procedure to compute $C_N(\mathbf{K})$ offers the possibility of significant savings of storage and computational costs, the factor which determines the actual savings that are obtained is how we choose to select the subnetworks σ since this choice directly determines the dimensionality of the convolution operations that are required in the course of obtaining $C_N(\mathbf{K})$. Since the computational and storage requirements involved in a convolution operation may be quantified, for any particular selection of subnetworks we may care to adopt, we may determine the overall space and time requirements *a priori*. Hence, in principle, it is possible to determine the subnetworks that will minimize these requirements. For large networks, however, this selection is a large combinatorial optimization problem. To solve this problem, Lam and Lien [LAM3] have proposed a heuristic optimization technique called *Tree Planting*. The Tree Convolution algorithm consists essentially of first finding an 'optimal' selection using Tree Planting and then computing $C_N(\mathbf{K})$ recursively using eq. 5.31. To compute the mean performance measures using eq. 5.6, it suffices to simply repeat the algorithm to obtain the normalization constants appearing in these formulae. As described in [LAM3], if certain intermediate results are stored in the course of computing $C_N(\mathbf{K})$, then it is not necessary that the entire algorithm be repeated.

REFERENCES

[AEI1] J.M. Aein, and O.S. Kosovych, Satellite Capacity Allocation, *IEEE Proceedings*, 65, 3, pp. 332-342, 1977.

[AKY1] I.F. Akyildiz, Mean Value Analysis for Blocking Queueing Networks, *IEEE Trans. on Software Eng.*, 14, 4, pp. 418-428, 1988.

[ALM1] G.T. Almes, and E.D. Lazowska, The Behaviour of Ethernet-Like Computer Communication Networks, in *Proc. 7th. ACM Symp. Oper. Syst. Princ.*, Asilomar, CA, pp. 66-81, 1979.

[BAL1] G. Balbo, and S.C. Bruell, Computational Aspects of Aggregation in Multiple Class Queueing Networks, *Performance Evaluation,* 3, pp. 177-185, 1983.

[BAL2] G. Balbo, S.C. Bruell, and S. Ghanta, The Solution of Homogeneous Queueing Networks with Many Job Classes, in *Proc. Int. Workshop on Modeling and Performance Evaluation of Parallel Systems*, pp. 385-417, Grenoble, France, 1984.

[BALS1] S. Balsamo, Decomposability in General Markovian Networks, in *Mathematical Computer Performance and Reliability*, G. Iazeolla, P.J. Courtois, and A. Hordijk (Eds.), pp. 3-13, Elsevier (North-Holland), 1984.

[BALS2] S. Balsamo, and G. Iazeolla, An Extension of Norton's Theorem for Queueing Networks, *IEEE Trans. on Software Eng.*, 8, 4, pp. 298-305, 1982.

[BAR1] A.D. Barbour, Networks of Queues and the Method of Stages, *Adv. Appl. Prob.*, 8, 3, pp. 584-591, 1976.

[BAS1] F. Baskett, K.M. Chandy, R.R. Muntz, and F. Palacios, Open, Closed and Mixed Networks of Queues with Different Classes of Customers, *J. ACM*, 22, pp. 248-260, 1975.

[BRA1] A. Brandwajn, A Model of a Time-Sharing System Solved Using Equivalence and Decomposition Methods, *Acta Informatica*, 4, 1, pp. 11-47, 1974.

[BRA2] A. Brandwajn, Fast Approximate Solution of Multiprogramming Models, *Performance Evaluation Review*, 11, pp. 141-149, 1982.

[BRA3] A. Brandwajn, Equivalence and Decomposition in Queueing Systems - A Unified Approach, *Performance Evaluation*, 5, pp. 175-186, 1985.

[BRU1] S.C. Bruell, and G. Balbo, *Computational Algorithms for Closed Queueing Networks*, North Holland, New York, 1980.

[BRU2] S.C. Bruell, G. Balbo, and P.V. Afshari, Mean Value Analysis of Mixed, Multiple Class BCMP Networks with Load Dependent Service Stations, *Performance Evaluation*, 4, pp. 241-260, 1984.

[BRY1] R.M. Bryant, A.E. Krzesinski, M.S. Lakshmi, and K.M. Chandy, The MVA Priority Approximation, *ACM Trans. on Computer Systems*, 2, 4, pp. 335-359, 1984.

[BUR1] D.Y. Burman, J.P. Lehoczky, and Y. Lim, Insensitivity of Blocking Probabilities in a Circuit-Switching Network, *J. Appl. Prob.*, 21, pp. 850-859, 1984.

[BUX1] W. Bux, Local-Area Subnetworks: A Performance Comparison, *IEEE Trans. on Computers*, 29, pp. 1465-1473, 1981.

[BUZ1] J.P. Buzen, Computational Algorithms for Closed Queueing Networks with Exponential Servers, *Commun. ACM*, 16, pp. 527-531, 1973.

[BUZ2] J.P. Buzen, Queueing Network Models of Multiprogramming, *Ph.D. Dissertation*, Div. Eng. Appl. Phys., Harvard University, Cambridge, Massachusetts, 1971.

[CAO1] W. Cao, and W.J. Stewart, Iterative Aggregation/Disaggregation for Nearly Uncoupled Markov Chains, *J. ACM*, 32, 3, pp. 702-719, 1985.

[CHA1] K.M. Chandy, U. Herzog, and L. Woo, Parametric Analysis of Queueing Networks, *IBM J. Res. and Dev.*, 19, pp. 36-42, 1975.

[CHA2] K.M. Chandy, and D. Neuse, Linearizer: A Heuristic Algorithm for Queueing Network Models of Computing Systems, *Commun. ACM*, 25, 2, pp. 126-134, 1982.

[CHA3] K.M. Chandy, and C.H. Sauer, Computational Algorithms for Product-Form Queueing Networks, *Commun. ACM*, 23, pp. 573-583, 1980.

[CHA4] K.M. Chandy, and C.H. Sauer, Approximate Methods for Analyzing Queueing Network Models of Computing Systems, *Computing Surveys*, 10, 3, pp. 281-317, 1978.

[CON1] A.E. Conway, Decomposition Methods and Computational Algorithms for Multiple Chain Closed Queueing Networks, *Ph.D. Dissertation*, Department of Electrical Engineering, University of Ottawa, Canada, 1986.

[CON2] A.E. Conway, A Polynomial Complexity Mean Value Analysis Algorithm for Multiple-Chain Closed Queueing Networks, in *Digest of Papers, IEEE Int. Symp. on Information Theory*, p. 61, Ann Arbor, Michigan, 1986.

[CON3] A.E. Conway, and N.D. Georganas, RECAL: A New Efficient Algorithm for the Exact Analysis of Multiple-Chain Closed Queuing Networks, *J. ACM*, 33, 4, pp. 768-791, 1986.

[CON4] A.E. Conway, and N.D. Georganas, Decomposition and Aggregation by Class in Closed Queueing Networks, *IEEE Trans. on Software Eng.*, 12, 10, pp. 1025-1040, 1986.

[CON5] A.E. Conway, and N.D. Georganas, A New Method for Computing the Normalization Constant of Multiple Chain Queueing Networks, *INFOR*, 24, 3, pp. 184-198, 1986.

[CON6] A.E. Conway, E. de Souza e Silva, and S.S. Lavenberg, Mean Value Analysis by Chain of Product Form Queueing Networks, *IEEE Trans. on Computers*, 38, 3, pp. 432-442, 1989.

[CON7] A.E. Conway, and N.D. Georganas, An Efficient Algorithm for Semi-Homogeneous Queueing Network Models, *Performance Evaluation Review*, 14, 1, pp. 92-99, 1986.

[CON8] A.E. Conway, The Solution of Local Area Distributed System Queueing Network Models with Several Customers in Each Routing Chain, in *Proc. 13th Biennial Symposium on Communications*, pp. B.2.1-B.2.4, Queen's University, Kingston, Canada, 1986.

[CON9] A.E. Conway, RECAL-MVA: A Polynomial Complexity Mean Value Analysis Algorithm for Multiple-Chain Closed Queueing Networks, unpublished work (available from the author upon request), December 1985.

[CON10] A.E. Conway, Product-Form and Insensitivity in Circuit-Switched Networks with Failing Links, *Performance Evaluation*, to appear.

[COO1] R. Cooper, *Introduction to Queueing Theory*, 2nd edition, North Holland, New York, 1981.

[COU1] P.J. Courtois, *Decomposability: Queueing and Computer System Applications,* Academic Press, New York, 1977.

[COU2] P.J. Courtois, Exact Aggregation in Queueing Networks, in *Proc. First Meeting AFCET-SMF on Applied Mathematics*, 1, pp. 35-51, Ecole Polytechnique, Palaiseau, France, 1978.

[COU3] P.J. Courtois, Error Analysis in Nearly Completely Decomposable Stochastic Systems, *Econometrica*, 43, pp. 691-709, 1975.

[COU4] P.J. Courtois, Error Minimization in Decomposable Stochastic Models, in *Applied Probability - Computer Science: The Interface*, 1, pp. 189-210, Birkhäuser, Stuttgart, 1982.

[COU5] P.J. Courtois, On Time and Space Decomposition of Complex Structures, *Commun. ACM*, 28, 6, pp. 590-603, 1985.

[COU6] P.J. Courtois, and P. Semal, Bounds for the Positive Eigenvectors of Nonnegative Matrices and Their Approximations by Decomposition, *J. ACM*, 31, 4, pp. 804-825, 1984.

[COU7] P.J. Courtois, and P. Semal, Computable Bounds for Conditional Steady-State Probabilities in Large Markov Chains and Queueing Models, *IEEE J. on Select. Areas in Communications*, 4, 6, pp. 926-937, 1986.

[COU8] P.J. Courtois, On the Near-Complete Decomposability of Networks of Queues and of Stochastic Models of Multiprogramming Computing Systems, *Sci. Rep.*, CMU-CS-72, 111, Carnegie-Mellon University, Pittsburgh, Pennsylvania, 1972.

[COX1] D.R. Cox, A Use of Complex Probabilities in the Theory of Stochastic Processes, *Proc. Cambridge Phil. Soc.*, 51, pp. 313-319, 1955.

[DAY1] J.D. Day, and H. Zimmermann, The OSI Reference Model, *Proc. IEEE*, 71, 12, pp. 1334-1340, 1983.

[DIJ1] N.M. van Dijk, On Jackson's Product Form with 'Jump-Over' Blocking, *Operations Research Letters*, 7, 5, pp. 233-235, 1988.

[DIS1] R.L. Disney, and D. König, Queueing Networks: A Survey of their Random Processes, *SIAM Review*, 27, 3, pp. 335-403, 1985.

[DOR1] J.B.M. van Doremalen, A Note on RECAL: the Recursion by Chain Algorithm, *Memorandum* COSOR 86-02, Eindhoven University of Technology, April 1986.

[DZI1] Z. Dziong, and J.W. Roberts, Congestion Probabilities in Circuit-Switched Integrated Services Network, *Performance Evaluation*, 7, pp. 267-284, 1987.

[EAG1] D.L. Eager, and K.C. Sevcik, Performance Bound Hierarchies for Queueing Networks, *ACM Trans. on Computer Systems*, 1, 2, pp. 99-115, 1983.

[EAG2] D.L. Eager, and K.C. Sevcik, Bound Hierarchies for Multiple-Class Queueing Networks, *J. ACM*, 33, 1, pp. 179-206, 1986.

[GEO1] N.D. Georganas, Modeling and Analysis of Message Switched Computer-Communication Networks with Multilevel Flow Control, *Computer Networks*, 4, pp. 285-294, 1980.

[GEO2] N.D. Georganas, Numerical Solution of Queueing Networks with Multiple Semiclosed Chains, *Proc. IEE*, 126, 3, pp. 229-231, 1979.

[GOL1] A. Goldberg, G. Popek, and S.S. Lavenberg, A Validated Distributed System Performance Model, in *Performance '83*, A.K.

Agrawala, and S.K. Tripathi (Eds.), pp. 251-268, North Holland, Amsterdam, 1983.

[GOR1] W.J. Gordon, and G.F. Newell, Closed Queueing Systems with Exponential Servers, *Operations Research*, 15, pp. 254-265, 1967.

[GOR2] W.J. Gordon, and G.F. Newell, Acknowledgement, *Operations Research*, p. 1182, 1967.

[GOY1] A. Goyal, S.S. Lavenberg, and K.S. Trivedi, Probabilistic Modeling of Computer System Availability, *Annals of Operations Research*, 8, pp. 285-306, 1987.

[GOY2] A. Goyal, and S.S. Lavenberg, Modeling and Analysis of Computer System Availability, *IBM J. Res. and Dev.*, 31, 6, pp. 651-664, 1987.

[GRE1] A.G. Greenberg, and J. McKenna, Solution of Closed, Product Form, Queueing Networks via the RECAL and Tree-RECAL Methods on a Shared Memory Multiprocessor, in *Proc. Int. Conf. on Measurement and Modeling of Computer Systems (Performance '89)*, Berkeley, CA, May 1989.

[HAM1] J.M. Hammersley, and D.C. Handscome, *Monte Carlo Methods*, Methuen, London, 1967.

[HAR1] P.G. Harrison, On Normalizing Constants in Queueing Networks, *Operations Research*, 33, 2, pp. 464-468, 1985.

[HEI1] P. Heidelberger, and S.S. Lavenberg, Computer Performance Evaluation Methodology, *IEEE Trans. on Computers*, 33, 12, pp. 1195-1220, 1984.

[HOY1] K.P. Hoyme, S.C. Bruell, P.V. Afshari, and R.Y. Kain, A Tree-Structured Mean Value Analysis Algorithm, *ACM Trans. on Computer Systems*, 4, 2, pp. 178-185, 1986.

[IRA1] K.B. Irani, and I.H. Öryüksel, A Closed Form Solution for the Performance Analysis of Multiple Bus Multiprocessor Systems, *IEEE Trans. on Computers*, 33, 11, pp. 1004-1012, 1984.

[IRL1] M.I. Irland, Buffer Management in a Packet Switch, *IEEE Trans. on Communications*, 26, 3, pp. 328-337, 1978.

[JAC1] J.R. Jackson, Networks of Waiting Lines, *Operations Research*, 5, pp. 518-521, 1957.

[JAC2] J.R. Jackson, Jobshop-Like Queueing Systems, *Management Science*, 10, 1, pp. 131-142, 1963.

[JACO1] P.A. Jacobson, and E.D. Lazowska, Analyzing Queueing Networks with Simultaneous Resource Possession, *Commun. ACM*, 25, pp. 142-151, 1982.

[JACO2] P.A. Jacobson, and E.D. Lazowska, A Reduction Technique for Evaluating Queueing Networks with Serialization Delays, in *Performance '83*, A.K. Agrawala and S.K. Tripathi (Eds.), pp. 45-59, North Holland, Amsterdam, 1983.

[KAM1] F. Kamoun, and L. Kleinrock, Analysis of Shared Finite Storage in a Computer Network Node Environment under General Traffic Conditions, *IEEE Trans. on Communications,* 28, 7, pp. 992-1003, 1980.

[KAU1] J.S. Kaufman, Blocking in a Shared Resource Environment, *IEEE Trans. on Communications*, 29, 10, pp. 1474-1481, 1981.

[KEL1] F.P. Kelly, Networks of Queues, *Adv. Appl. Prob.*, 8, pp. 416-432, 1976.

[KEL2] F.P. Kelly, Networks of Queues with Customers of Different Types, *J. Appl. Prob.,* 12, pp. 542-554, 1975.

[KEL3] F.P. Kelly, *Reversibility and Stochastic Networks*, Wiley, New York, 1980.

[KIN1] P.J.B. King, and I. Mitrani, Modeling the Cambridge Ring, in *Proc. ACM SIGMETRICS Conf. Meas. Model. Comp. Syst.*, pp. 250-258, Seattle, Washington, 1982.

[KING1] J.F.C. Kingman, Markov Population Processes, *J. Appl. Prob.*, 6, pp. 1-18, 1969.

[KLE1] L. Kleinrock, *Communication Nets - Stochastic Message Flow and Delays*, McGraw-Hill, New York, 1964.

[KOE1] E. Koenigsberg, Cyclic Queues, *Opnl. Res. Q.*, 9, pp. 22-35, 1958.

[KOE2] E. Koenigsberg, Comments on "On Normalizing Constants in Queueing Networks" by P.G. Harrison, *Operations Research*, 34, 2, p. 330, 1986.

[KRI1] P.S. Kritzinger, A Performance Model of the OSI Communication Architecture, *IEEE Trans. on Communications*, 34, 6, pp. 554-563, 1986.

[KRI2] P.S. Kritzinger, S. van Wyk, and A.E. Krzesinski, A Generalization of Norton's Theorem for Multiclass Queueing Networks, *Performance Evaluation*, 2, pp. 98-107, 1982.

[KRZ1] A.E. Krzesinski, Multiclass Queueing Networks with State-Dependent Routing, *Performance Evaluation*, 7, pp. 125-143, 1987.

[KRZ2] A.E. Krzesinski, Multiclass Queueing Networks with State-Dependent Routing, *IBM Research Report*, RC-9761, Yorktown Heights, New York, 1982.

[KRZ3] A. Krzesinski, and P. Teunissen, Multiclass Queueing Networks with Population Constrained Subnetworks, *Performance Evaluation Review*, 13, 2, pp. 128-139, 1985.

[KUR1] J.F. Kurose, and H.T. Mouftah, Computer-Aided Modeling, Analysis, and Design of Communication Networks, *IEEE J. on Select. Areas in Communications*, 6, 1, pp. 130-145, 1988.

[LAM1] S.S. Lam, Dynamic Scaling and Growth Behaviour of Queueing Network Normalization Constants, *J. ACM*, 29, pp. 492-513, 1982.

[LAM2] S.S. Lam, A Simple Derivation of the MVA and LBANC Algorithms from the Convolution Algorithm, *IEEE Trans. on Computers*, 32, 11, pp. 1062-1064, 1983.

[LAM3] S.S. Lam, and Y.L. Lien, A Tree Convolution Algorithm for the Solution of Queueing Networks, *Commun. ACM,* 26, pp. 203-215, 1983.

[LAM4] S.S. Lam, Queueing Networks with Population Size Constraints, *IBM J. Res. and Dev.*, 21, pp. 370-378, 1977.

[LAM5] S.S. Lam, and J.W. Wong, Queueing Network Models of Packet Switching Networks, Part 2: Networks with Population Size Constraints, *Performance Evaluation*, 2, pp. 161-180, 1982.

[LAM6] S.S. Lam, Carrier Sense Multiple Access Protocol for Local Networks, *Computer Networks*, 4, pp. 21-32, 1980.

[LAV1] S.S. Lavenberg, and M. Reiser, Stationary State Probabilities at Arrival Instants for Closed Queueing Networks with Multiple Types of Customers, *J. Appl. Prob.*, 17, pp. 1048-1061, 1980.

[LAV2] S.S. Lavenberg (Ed.), *Computer Performance Modeling Handbook*, Academic Press, New York, 1983.

[LAZ1] E.D. Lazowska, J. Zahorjan, and K.C. Sevcik, Computer System Performance Evaluation using Queueing Network Models, *Ann. Rev. Comput. Sci.*, 1, pp. 107-137, 1986.

[LAZ2] E.D. Lazowska, and J. Zahorjan, Multiple Class Memory Constrained Queueing Networks, *Performance Evaluation Review*, 11, pp. 130-140, 1982.

[LEB1] J. Le Boudec, A BCMP Extension to Multiserver Stations with Concurrent Classes of Customers, *Performance Evaluation Review*, 14, 1, pp. 78-91, 1986.

[LEB2] J. Le Boudec, The MULTIBUS Algorithm, *Performance Evaluation*, 8, pp. 1-18, 1988.

[LEM1] A.J. Lemoine, Networks of Queues - A Survey of Equilibrium Analysis, *Management Science*, 24, 4, pp. 464-481, 1977.

[LIT1] J.D.C. Little, A Proof of the Queueing Formula $L = \lambda W$, *Operations Research*, 9, pp. 383-387, 1961.

[MAR1] R.A. Marie, An Approximate Analytical Method for General Queueing Networks, *IEEE Trans. on Software Eng.*, 5, 5, pp. 530-538, 1979.

[MARS1] M.A. Marsan, G. Balbo, G. Chiola, and S. Donatelli, On the Product Form Solution of a Class of Multiple Bus Multiprocessor System Models, in *Proc. Int. Workshop on Modeling and Performance*

Evaluation of Parallel Systems, Grenoble, France, 1984. Also in *J. Syst. Software*, 6, pp. 117-124, 1986.

[MARS2] M.A. Marsan, G. Balbo, and G. Conte, *Performance Models of Multiprocessor Systems*, The MIT Press, Cambridge, Massachusetts, 1986.

[MCK1] J. McKenna, and D. Mitra, Integral Representations and Asymptotic Expansions for Closed Markovian Queueing Networks: Normal Usage, *BSTJ*, 61, 5, pp. 661-683, 1982.

[MCK2] J. McKenna, Extensions and Applications of RECAL in the Solution of Closed Product Form Queueing Networks, *Communications in Statistics, Stochastic Models*, 4, 2, pp. 235-276, 1988.

[MCK3] J. McKenna, *private communication*, 1988.

[MCK4] J. McKenna, D. Mitra, and K.G. Ramakrishnan, A Class of Closed Markovian Queueing Networks: Integral Representations, Asymptotic Expansions, Generalizations, *BSTJ*, 60, 5, pp. 599-641, 1981.

[MOO1] F.R. Moore, Computational Model of a Closed Queueing Network with Exponential Servers, *IBM J. Res. and Dev.*, 16, pp. 567-572, 1972.

[MUN1] R.R. Muntz, Queueing Networks: A Critique of the State of the Art and Directions for the Future, *Computing Surveys*, 10, 3, pp. 353-359, 1978.

[NEU1] D.M. Neuse, Approximate Analysis of Large and General Queueing Networks, *Ph.D. Dissertation*, The University of Texas at Austin, 1982.

[NEU2] D.M. Neuse, and K.M. Chandy, HAM: The Heuristic Aggregation Method for Solving General Closed Queueing Network Models of Computing Systems, *Performance Evaluation Review*, 11, pp. 99-112, 1982.

[OSI1] *Data Communication Networks Open Systems Interconnection (OSI) System Description Techniques*, Red Book, Vol. 8, Fasc. 8.5, CCITT, Geneva, 1985.

[PEN1] M.C. Pennotti, and M. Schwartz, Congestion Control in Store and Forward Tandem Links, *IEEE Trans. on Communications*, 23, 12, pp. 1434-1443, 1975.

[PIT1] B. Pittel, Closed Exponential Networks of Queues with Saturation: The Jackson Type Stationary Distribution and its Asymptotic Analysis, *Math. Oper. Res.*, 4, 4, pp. 367-378, 1979.

[RAB1] L.R. Rabiner, and B. Gold, *Theory and Applications of Digital Signal Processing*, Prentice-Hall, Englewood Cliffs, New Jersey, 1975.

[RAM1] K.G. Ramakrishnan, and D. Mitra, An Overview of PANACEA, a Software Package for Analyzing Markovian Queueing Networks, *BSTJ*, 61, 10, pp. 2849-2872, 1982.

[REI1] M. Reiser, A Queueing Network Analysis of Computer Communication Networks with Window Flow Control, *IEEE Trans. on Communications*, 27, pp. 1199-1209, 1979.

[REI2] M. Reiser, Performance Evaluation of Data Communication Systems, *IEEE Proceedings*, 70, 2, pp. 171-196, 1982.

[REI3] M. Reiser, and H. Kobayashi, Queueing Networks with Multiple Closed Chains: Theory and Computational Algorithms, *IBM J. Res. and Dev.*, 19, pp. 283-294, 1975.

[REI4] M. Reiser, and S.S. Lavenberg, Mean Value Analysis of Closed Multichain Queueing Networks, *J. ACM*, 27, pp. 313-322, 1980.

[REI5] M. Reiser, Communication-System Models Embedded in the OSI-Reference Model: A Survey, in *Computer Networking and Performance Evaluation*, T. Hasegawa, H. Takagi, and Y. Takahashi (Eds.), pp. 85-111, Elsevier (North-Holland), 1986.

[REI6] M. Reiser, Admission Delays on Virtual Routes with Window Flow Control, in *Performance of Data Communication Systems and their Applications*, G. Pujolle (Ed.), pp. 67-76, North Holland, Amsterdam, 1981.

[REI7] M. Reiser, and C.H. Sauer, Queueing Network Models: Methods of Solution and their Program Implementation, *IBM Research Report*, RC 6109, Yorktown Heights, New York, 1976.

[REI8] M. Reiser, and H. Kobayashi, Recursive Algorithms for General Queueing Networks with Exponential Servers, *IBM Research Report*, RC 4254, Yorktown Heights, New York, 1973.

[REI9] M. Reiser, Mean-Value Analysis and Convolution Method for Queue-Dependent Servers in Closed Queueing Networks, *Performance Evaluation*, 1, pp. 7-18, 1981.

[SAU1] C.H. Sauer, Computational Algorithms for State-Dependent Queueing Networks, *ACM Trans. on Computer Systems*, 1, 1, pp. 67-92, 1983.

[SAU2] C.H. Sauer, Corrigendum: Computational Algorithms for State-Dependent Queueing Networks, *ACM Trans. on Computer Systems*, 1, 4, p. 369, 1983.

[SCH1] M. Schwartz, Performance Analysis of the SNA Virtual Route Pacing Control, *IEEE Trans. on Communications*, 30, 1, pp. 172-184, 1982.

[SEV1] K.C. Sevcik, and I. Mitrani, The Distribution of Queueing Network States at Input and Output Instants, *J. ACM*, 28, pp. 358-371, 1981.

[SIM1] H.A. Simon, and A. Ando, Aggregation of Variables in Dynamic Systems, *Econometrica*, 29, pp. 111-138, 1961.

[SOU1] E. de Souza e Silva, and S.S. Lavenberg, A Mean Value Analysis by Chain Algorithm for Product Form Queueing Networks, *IBM Research Report*, RC 11641, Yorktown Heights, New York, 1986.

[SOU2] E. de Souza e Silva, and S.S. Lavenberg, Calculating Joint Queue-Length Distributions in Product-Form Queueing Networks, *J. ACM*, 36, 1, pp. 194-207, 1989.

[SOU3] E. de Souza e Silva, S.S. Lavenberg, and R.R. Muntz, A Clustering Approximation Technique for Queueing Network Models with a Large Number of Chains, *IEEE Trans. on Computers*, 35, 5, pp. 419-430, 1986.

[SOU4] E. de Souza e Silva, and R.R. Muntz, Approximate Solutions for a Class of Non-Product Form Queueing Network Models, *Performance Evaluation*, 7, pp. 221-242, 1987.

[SOU5] E. de Souza e Silva, S.S. Lavenberg, and R.R. Muntz, A Perspective on Iterative Methods for the Approximate Analysis of Closed Queueing Networks, in *Mathematical Computer Performance and Reliability*, G. Iazeolla, P.J. Courtois, and A. Hordijk (Eds.), Elsevier Science Publishers B.V. (North-Holland), 1984.

[SPR1] J. Spragins, Analytical Queueing Models: Guest Editor's Introduction, *IEEE Computer*, 13, 4, pp. 9-11, 1980.

[STA1] W. Stallings, *Handbook of Computer Communications Standards: Volume 1*, Macmillan, New York, 1987.

[STE1] W.J. Stewart, A Comparison of Numerical Techniques in Markov Modeling, *Commun. ACM*, 21, pp. 144-151, 1978.

[SUR1] R. Suri, Robustness of Queueing Network Formulas, *J. ACM*, 30, 3, pp. 564-594, 1983.

[SWE1] R.J. Swersey, Some Extension of Cyclic Queue Theory, Operations Research Center ORC 65-14, College of Engineering, University of California, Berkeley, 1965. (Also Closed Networks of Queues ORC 67-1, 1967.)

[TOW1] D. Towsley, Queueing Network Models with State-Dependent Routing, *J. ACM*, 27, 2, pp. 323-337, 1980.

[TOW2] D. Towsley, Local Balance Models of Computer Systems, Tech. Rep. TR-60, Dept. Comp. Sci., University of Texas, Austin, 1975.

[TUC1] S. Tucci, and C.H. Sauer, The Tree MVA Algorithm, *Performance Evaluation*, 5, pp. 187-196, 1985.

[VAN1] H. Vantilborgh, Exact Aggregation in Exponential Queueing Networks, *J.ACM*, 25, 4, pp. 620-629, 1978.

[VAN2] H. Vantilborgh, R.L. Garner, and E.D. Lazowska, Near-Complete Decomposability of Queueing Networks with Clusters of Strongly Interacting Servers, *Performance Evaluation Review*, 9, pp. 81-92, 1980.

[WON1] J.W. Wong, Queueing Network Modeling of Computer Communication Networks, *Computing Surveys*, 10, 3, pp. 343-351, 1978.

[WON2] J.W. Wong, and S.S. Lam, Queueing Network Models of Packet Switching Networks, Part 1: Open Networks, *Performance Evaluation*, 2, pp. 9-21, 1982.

[YAO1] D.D. Yao, and J.A. Buzacott, Modeling a Class of Flexible Manufacturing Systems with Reversible Routing, *Operations Research*, 35, 1, pp. 87-93, 1987.

[ZAH1] J. Zahorjan, E.D. Lazowska, and R.L. Garner, A Decomposition Approach to Modeling High Service Time Variability, *Performance Evaluation*, 3, pp. 35-54, 1983.

INDEX

A

B

C

D

H

I

J

K

L

M

N

O

P

Q

R

T

U

V

W

Z

The MIT Press, with Peter Denning, general consulting editor, and Brian Randell, European consulting editor, publishes computer science books in the following series:

ACM Doctoral Dissertation Award and Distinguished Dissertation Series

Artificial Intelligence
Patrick Winston, founding editor
Michael Brady, Daniel Bobrow, and Randall Davis, editors

Charles Babbage Institute Reprint Series for the History of Computing
Martin Campbell-Kelly, editor

Computer Systems
Herb Schwetman, editor

Explorations with Logo
E. Paul Goldenberg, editor

Foundations of Computing
Michael Garey and Albert Meyer, editors

History of Computing
I. Bernard Cohen and William Aspray, editors

Information Systems
Michael Lesk, editor

Logic Programming
Ehud Shapiro, editor
Fernando Pereira, Koichi Furukawa, and David H. D. Warren, associate editors

The MIT Press Electrical Engineering and Computer Science Series

Scientific Computation
Dennis Gannon, editor

Technical Communication
Ed Barrett, editor

Into the Light

Poems and Short Stories by

Linda Menzies

ISBN number 978 0 902303 79 9

Published by the Workers' Educational Association, Riddle's Court, 322 Lawnmarket, Edinburgh, EH1 2PG. The Workers' Educational Association is a charity registered in England and Wales (number 1112775) and in Scotland (number SC039239) and a company limited by guarantee registered in England and Wales (number 2806910). Registered address is WEA, 4 Luke Street, London, EC2A 4XW.

Printed and bound in Great Britain by
Pandaprint, 104 Park Road, Rosyth KY11 2JL